SIMULTANEOUS ENGINEERING FOR NEW PRODUCT DEVELOPMENT

SIMULTANEOUS ENGINEERING FOR NEW PRODUCT DEVELOPMENT

Manufacturing Applications

JACK RIBBENS
New Product Development
Inverness, IL

JOHN WILEY & SONS, INC.
New York • Chichester • Weinheim • Brisbane • Singapore • Toronto

Copyright © 2000 by John Wiley & Sons, Inc. All rights reserved.

Published simultaneously in Canada.

This publication is designed to provide accurate and authoritative information in regard to the subject matter covered. It is sold with the understanding that the publisher is not engaged in rendering professional services. If professional advice or other expert assistance is required, the services of a competent professional person should be sought.

Library of Congress Cataloging-in-Publication Data:

Ribbens, Jack
   Simultaneous engineering for new product development : manufacturing applications / by Jack Ribbens.
      p.   cm.
   ISBN 0-471-25265-4 (cloth : alk. paper)
   1. New products.   2. Concurrent engineering.   3. Computer integrated manufacturing systems.   4. Product engineering—Case studies.   I. Title.
   TS170.R53   2000
   658.5'75—dc21   99-046112

Printed in the United States of America.

10 9 8 7 6 5 4 3 2 1

FOREWORD

As a Chinese philosopher once said, "May you live in interesting times." In fact, it is clear that today we are living in interesting times, very interesting times, and challenging as well. Certainly, this notion applies to manufacturing and the product development process that underlies manufactured products. While my expertise is primarily focused on the automotive industry, it is evident that much of what we are witnessing in this industry applies to other manufacturing industries as well. At the most fundamental level, there are remarkable similarities when we consider such factors as these:

- Exploding knowledge, including technology
- Dissolving global boundaries and the acceleration of globalization as companies develop globally integrated organizations
- More demanding customers for all goods and services
- Changing industry structure, including mergers and acquisitions, and a new order of the manufacturer/supplier relationship, as suppliers take increased responsibility for the design of parts, modules, and systems
- Simultaneous engineering, which, at its most rudimentary level, brings product engineering and manufacturing together
- A profound and rapidly growing role for information technology in all of its forms
- Intensifying competition
- Human resource challenges as all of industry becomes more skill dependent, even as retirements accelerate
- Time pressures that demand more, better, yet faster results

- Dynamic leadership committed to effectively managing change and serving as coach of the team

It is in this context that product development is taking center stage in entire manufacturing industries and within individual companies. Never before have we seen such great emphasis on, and need for, effective product development processes. It is also clear that product development, in the current era, is far more than sketch, engineer, and build. Today, it is a time-driven, knowledge-based process that involves every aspect of the enterprise from market assessment and initial conceptualization to the selling process and tracking throughout the full product life cycle. It should be a highly integrated and well-defined process that accords proper attention to every element. It can no longer be an ad hoc process, but rather a process involving very specific steps and methods. This does not mean that it need be highly rigid or that there is only one "right" process. In fact, there are many effective product development solutions to the same problem, but all the bases must be touched. In a given organization, there is no longer room for freelance product development; it must adhere to the company's specific process.

Jack Ribbens has written an excellent book on product development with specific emphasis on the automotive, aerospace, electronic, and defense industries. It is generally applicable, however, to any manufactured product. The essential rules and methods remain the same. Jack has identified the key steps in the product development process and has applied them to specific industrial cases. While he addresses specific industries and products, he also creates a broader context for the overall process as well.

This is certainly a book for young, inexperienced engineers, designers, and technicians; it could readily be used in colleges, universities, and community colleges, as well as serving as a training guide in industrial organizations. It also has application for experienced practitioners, particularly when we consider the massive level of retraining being done with existing employees in response to a fast-changing competitive environment. Finally, I believe *Simultaneous Engineering for New Product Development: Manufacturing Applications* certainly has considerable value from a remedial or updating viewpoint. In general, it could help bring an organization's thinking back onto the proper track.

Today, there is no room for multiple processes and an ad hoc methodology within any given organization. It is absolutely necessary to have a common process, although as noted earlier, this can differ in detail rather significantly from one organization to another. Certainly, this book can assist in developing a robust process that is a fit with both the culture of the organization and the rapidly evolving new world.

Clearly, technology is having a profound and growing impact on the product development process. Math-based simulation tools are promising to significantly reduce the time and cost of many facets of the process, from the initial styling and conceptualization phases through engineering (involving the use of virtual prototypes) and manufacturing. Furthermore, information technology tools are becoming increasingly critical to the communication and coordination of the total process.

For those of us with an engineering background, the example problems common to any engineering text are well presented by the case studies in Jack Ribbens's book. These are very helpful in reducing the abstraction of the process. Ultimately, product development is not an exercise in creating an imaginary or virtual product. It must be aimed at developing a real product that people and organizations want to buy at an affordable price. I am enthusiastic about this book, and believe it will be of considerable value to students and practitioners alike.

DAVID E. COLE

Office for the Study of Automotive Transportation
University of Michigan Transportation Research Institute

PREFACE

I have been interested in Simultaneous Engineering for the last 20 years, when I worked for one of the big three automobile companies in the United States. I was in one of the engineering staff groups assigned to monitor important design activities for key parameters, including weights, warranty expense, product cost planning, prototype development, among others. The staff would meet every day for several hours on a certain topic, with the appropriate release group. Monday was cost reduction, Tuesday was weights, Wednesday warranty reduction, Thursday standardization and simplification, and so on. New engineering proposals were debated, product-change notices were written, and all manner of business was conducted in these interminable meetings. At the end of the week, the staff had to report to senior management the status of all the programs and the variances to target for weights, costs, and so forth. However, it was all guesswork because, while everyone agreed that the variables being debated were highly interactive, no one had sufficient information or resources to accurately predict the joint outcomes for all of the programs and changes in process. I began to think that a prediction model would be a worthwhile endeavor for study, which I began as part of my MBA program (albeit informally) but never finished.

Later, after I began working in the insurance industry, I had more opportunities to continue the process as we started reviewing new vehicle programs for damageability and repairability problems. It became obvious during our discussions with automotive engineers that the interactivity between design variables was extremely complex and occurred only infrequently with much priority. When an interaction potential was raised—such as the mass and manufacturing complexity implications of providing factory seams adjacent to crush zones—for ease of partial replacement, there was usually an uncomfortable silence before they tried to redirect the discussion back to the repairability of plastic bumper covers and similar components.

That was then. Now, 20 years later, auto manufacturers have almost entirely computerized their new product design and engineering processes, utilizing programs such as Catia, and others. However, the basic issue remains: many engineers and designers do not have the time or experience to get involved in functions outside of their immediate area, unless specifically required for interface development, for mating part tolerances, clearances for assembly or system compatibility. It is rightfully senior management's reponsibility to insure that all program parameters are being met or exceeded. However, I believe that designers, engineers, and technicians can contribute more to their companies and careers if they have a heightened awareness of the potential interactions between design parameters and how theoretical solutions can provide practical solutions by thinking through the problems that transcend their individual engineering assignments. To this end, the book is hereby dedicated.

I wanted to recognize the assistance that I have been given over the past few years in doing the research completing this text. Valuable insight and constructive criticism are very important to new writers, and I received a great deal of both along with support from the following:

Paul Shefferly	DaimlerChrysler Corporation
Robert Dubensky	DaimlerChrysler Corporation
Vince Peter Render	Ford Motor Company
Frank Wassilak	Ford Motor Company
Larry Pecar	General Motors Corporation
Ian Findlay	General Motors Corporation
Jerry Frank	3M Company
John Kushner	3M Company
Dr. Geoffrey Boothroyd	URI/IME & Boothroyd-Dewhurst Inc.
Dr. Peter Dewhurst	URI/IME & Boothroyd-Dewhurst Inc.
Alfredo Herrera	McDonnell-Douglas Division of The Boeing Company
Naomi Bulock	The Mathworks. Inc.
Danielle Rabina	The Mathworks Inc.
Shelley Marak	The Mathworks Inc.
Elaine Person	The Mathworks Inc.
Dieter Anselm	Allianz-Zentrum für Technik GmbH

Paris Gogos	EMS Corporation
Judy Vaughn	EMS Corporation
Cathy Rowe	OSAT/UMTRI
Dr. David E. Cole	OSAT/UMTRI
Joseph Kormos	Precision Devices, Inc.
Debbie Notar	Precision Devices, Inc.

I want to give a special thanks to my brother and sister-in-law, William and Catherine, for providing guidance with math problems and editing, respectively. Finally, my warmest gratitude and love go to my wife, Barbara, for supporting me so faithfully in all of this for so long.

CONTENTS

SECTION I
NEW PRODUCT DEVELOPMENT PROCESS PHASE ELEMENTS

This unit contains separate chapters on each of the new product development process phase elements discussed in the introduction. While it is important that the reader understand all phases of new product development, in order to facilitate the learning process, the principles of Simultaneous Engineering also need to be applied at each phase of the new product development process. This should enable the reader to differentiate between superior and inferior new product concepts and to proactively participate in decision making regarding each phase.

It is also important to remember that, while expertise in the individual areas typically takes years to acquire, the fundamental knowledge, contained herein, should provide the reader with sufficient confidence to participate actively in any new product development discussions and analysis. Learning the concepts by performing the tasks is a basic premise of this book.

Although the new product development process phase elements are presented and analyzed sequentially, this should not limit the development of each element simultaneously, as much as possible.

Although this is not a text on how to design products, it is concerned with identifying all of the elements that must be considered when new products are being created, as early in the concept stage as possible and also at each phase element of the new product development process. The subsequent analysis and interaction of each parameter will show how the design parameters, acting together, could positively influence the total product performance.

1

STRATEGIC PLANNING

1.1 OVERALL PROCESS DESCRIPTION AND WHY IT IS NEEDED

Strategic planning is the foundation for managing the future of any business, including manufactured products, for which there are unique requirements due to the need to procure long lead items such as tooling and assembly fixtures. Simultaneous Engineering needs to have a special and prominent place even in this early planning phase, in order to insure that all issues that could affect the new product have been considered and acted upon. The process begins with asking several questions regarding key issues, some of which are as follows:

Does the company have a formal strategic direction that is shared with all concerned, i.e., associates (employees), community, customers, investors and suppliers?

Does the strategy cover all industries, markets, and products that the company touches?

Where will the business need to be in 2 years, 5 years, 10 years?

Where will the competition be in 2 years, 5 years, 10 years?

Do all of the existing product/market strategies fit together, to maximize both short- and long-term profitability and growth of the company?

What are the new product development strategies, and how will they affect the outcomes planned for the business?

How well do they fit with existing products or businesses?

These questions and their answers will generate more questions as the process moves along. The foremost strategic decision has to do with planning for activities which will sustain profitable growth for the firm, balanced, simultaneously, at several levels, between the following factors:

Short- and long-term financial requirements

Product creativity and cost control

Advanced design concepts and manufacturing capabilities

Human resources

Development time to market

Product compliance

Addressing customer needs and concerns

Questions regarding how important issues such as the interaction of the above will be addressed in more detail in later chapters.

New product development is one path that a company might take to provide for future profitable growth. Many firms develop their own products internally, with input from workers, customers, suppliers, and outside consultants. Some companies choose to obtain new technology through mergers with, and acquisitions of, other companies. The trade-off between internal and external new product development is also a balance of the following combination of factors:

Development cost

Total time to market

Need for security

Level of innovation required for success (to compete), in the market.

All of the above factors make a case for a disciplined approach to new product development that is supported and facilitated by senior management; that benefits from the experience of new product development teams comprised of representatives from all affected activities; and that performs, within clearly delineated phase steps, extremely well-defined tasks for improved understanding, greater urgency, and reduced risks of failure. Simultaneous Engineering fits precisely into this environment, at each phase of the new product development process.

Balancing the need to perform new product development rapidly is the requirement that it be done "right." The historic rate of new product acceptance in the market (i.e., achieving sufficient earning power to contribute significantly to the corporate profitability) has been abysmally low over the last few decades. The requirement to plan, execute, and follow up is imperative for new product success. Specifically, "doing it right" means meeting all customer, legal, regulatory, functional performance, and marketing requirements, without having to redesign the product for manufacturing, sales, service, or any other consideration. It also involves thinking through each step prior to actually performing the step. Missing elements should be gathered or analyzed before beginning each new step.

The balance theme generates the following questions to be answered, through extensive analyses within the company, along with the involvement of key customers and suppliers, if possible, in order to prioritize new product development efforts.

Once it has been established that new product development activity would enhance the firm's profitability, answers to the following questions further define how much effect the new product would need to have on the existing product lines:

What are the sales volume expectations?

What are the market share expectations?

What are the gross profit and breakeven volume expectations?

What is the cannibalization potential from other company product lines?

What competitive countermoves are anticipated?

What is the projected new product life cycle?

Answering these questions in an objective and consistent manner should provide the right directions for successful new product programs. Detailed Simultaneous Engineering plans and procedures for each phase of the new product development process will be explained later in the text. Because of the importance of manufactured products in a global economy, this book will concentrate on developing strategies that utilize advanced analytical techniques such as parametric design modeling to streamline the new product development process, one which is not, at this point, known for its speed or accuracy.

1.2 HOW NEW PRODUCTS FIT WITH EXISTING PRODUCT LIFE CYCLES

Many manufacturing businesses have existing products that have reached or are just reaching their maturity in the market. These "cash cows" are successful enough to pay for ongoing operations, having also paid back their development costs, and are now in a position to generate sufficient cash for their replacement or for an entirely new product. The challenge is to create a new product that not only maintains the current product life cycle but can be adapted to compensate for any future product life-cycle changes, particularly where they are becoming foreshortened. Consumer goods and some electronic products seem to be following this path.

Other products have longer product life cycles, particularly those with commercial applications, such as airliners and construction equipment and machinery, where 20-year life cycles are not uncommon since they are based on regularly scheduled maintenance and key component replacement programs. On the other hand, in the automobile industry, product development time lines are running about 3 years, just ahead of the product's primary, first-owner life cycle of 5 years. This includes leased as well as purchased vehicles. While most vehicles purchased new are retained approximately 8–10 years by their first owner, lease turnover is much shorter, resulting in a composite life cycle of approximately 5 years.

The management requirements for mature products are far different from those of new products. Budgets, expenditures, research, and further development of the existing product have to be strictly controlled, since its life-cycle position would indicate that further expenditures may not result in increased sales. However, new products need a different management approach, one that will evaluate the risks of success and failure, investigate new opportunities, launch new ventures, nurture new concepts, and sustain development of new products, often in a difficult and hostile environment, that is, the existing product management. It takes strong leadership to institute a revolutionary policy of this magnitude, while not alienating the people involved with everyday operations.

New product development may have to be done at a different site than the same laboratory where regular production samples are audited. Current production problems are sometimes difficult to ignore when one is constantly bombarded by everyday difficulties with suppliers, customers, unions, and so on. Managers of regular production are striving for stability and consistency as indicators of the highest quality, while entrepreneurs seek just the opposite: product innovation through unsystematic questioning of the status quo and striving for change to improve product performance and innovation.

Senior management has to appoint an officer or at least a director with sufficient authority and span of control to provide the leadership necessary to meet the demands of new product development schedules. The dedication of the representatives from each affected area should be unconditional, so that crucial issues of the project will have their undivided attention and energy.

The critical factors that were so vital to the success of the existing product may have little or nothing to contribute toward a new one. A completely different attitude and line of reasoning may be needed before a new product can be approached, even within the same company. A fresh attitude is frequently necessary in order to gain a different perspective on what the new product needs to accomplish, often in a drastically changing market. Existing product management quite often cannot and should not be put in charge of its replacement. All too often, when this happens, it is easy to divert resources, originally dedicated to the new effort, toward solving some of the existing product faults, thereby giving the latter a new lease on life, albeit a short one.

1.3 IDENTIFYING NEW PRODUCT OPPORTUNITIES

New product opportunities can be identified from a number of sources, including competitive intelligence, in-house research facilities, and outside consulting firms. Deciding which opportunity is the best is the result of extensive analysis and accurately assessing the risks and rewards of the decision. External factors could also influence any data analysis. Economic conditions, political actions, or even the weather could play a significant role for a particular product. Most of the above activities may identify trends, outcomes, or predictions, based on their own or industry data, that may provide the stimulus for further study.

For example, it was once generally accepted in the North American automobile industry that small cars could not be manufactured and marketed successfully, because Asian manufacturers held an insurmountable cost advantage. The fact that one manufacturer not only questioned this hypothesis but also developed a new vehicle concept to compete in this segment shows that any proposal, no matter how inconceivable at the time, can be considered viable.

How senior management chooses to identify and deal with the above phenomena has everything to do with how new product development is managed. If newfound fortune or experience is treated casually or arrogantly, then whatever gains there are may indeed be short term only. If, however, a regular review and analysis process is in place to exploit new opportunities, regardless from where the occurrences originate, then the new product development process can be maximized.

Examples of these situations are illustrated in the case studies that follow in later chapters.

In summary, the ideal organization to tackle new product development should include senior management and/or representatives from all areas of the company, key customers and suppliers, and other interested stakeholders, if applicable. For the sake of convenience, this group will be referred to from now on as the "new product committee." While the product planning department typically coordinates this activity, the opportunity for other interested employees to contribute, needs to be provided. Leadership of the new product committee is frequently referred to as the "product champion," one who shows a personal as well as a professional attitude and dedication toward the new product program. The background and training requirements for a product champion can vary widely; however, engineers with good business sense seem quite often to have the right combination of qualifications.

1.4   FORMULATING THE RIGHT STRATEGY

Once a company decides that a new product program is necessary, how should it go about organizing and planning for this effort? Even if the firm already has a well-developed new business strategy, how it gets interpreted and implemented within the organization can have a major impact on its success.

Many factors can contribute to the need to reevaluate a corporate business strategy. Changes in market conditions, customer wants, financial performance, competitive initiatives, government regulations, and so forth are all valid reasons why business strategies need to be somewhat flexible. The most important outcome is to provide for a formal reevaluation schedule, that is, once a quarter, yearly, or similar. Monitoring of all pertinent activities, applicable to the review process, also need to be linked to the same schedule. The factors and their measurements have to be analyzed together, with priorities established and weighted properly, so that a fact-based comparison of results is possible, another scenario for Simultaneous Engineering.

The analysis of available information requires more than a basic understanding of economic theory. It requires unique expertise that is aggressively sought and highly compensated by many large corporations. However, this text is intended to provide sufficient knowledge to permit a rudimentary analysis for decision making based on realistic facts and figures obtained from the business itself or industry sources. The key issue is the relationship between the management of new product development, and how much input to it is sought from all of the potential information sources available, both inside and outside the company. The manufacturers that tend to do everything themselves have to manage all of the marketing and financial analyses inside the firm and develop strategies that address all of the global competitiveness issues adequately, and on a timely basis. Other companies utilize outside consulting firms to assist them in achieving cost-effective and timely results, particularly if the expertise does not reside within the company.

This is not an intentional oversimplification of a very complex process. There are many ways to manage innovation, and the right method, if there is one, depends on the circumstances occurring at the moment, at an individual company. The key is to develop a strategy that will unite and energize the firm and grow the expertise for its own new product development efforts.

Within the marketing arena, new product development is typically counted on to provide access to new customers and markets, to increase revenues and profits, and to carry forward the company's vision to both the buying and investing public. There are several avenues by which the firm can pursue such a strategy. Extending current products into new markets and upgrading current products are usually the easiest and cheapest method of reaching new customers. However, these actions typically have a shorter life cycle than the original product, particularly if the competition is actively pursuing a completely new product strategy.

The role to be played by Simultaneous Engineering is a continuous review of benchmarked capabilities, a matrix developed for each product versus its direct competitors, for each significant attribute. If discussions of potential new products cross over the boundaries of feasibility, cost, marketability, and so on, responsible managers have to interrupt the discussion and advise against further analysis. On the other hand, if the discussions do not go far enough to result in a new product that is sufficiently revolutionary or at least evolutionary, the same managers need to advocate for a higher level of innovation. The checklists for the first 10 chapters in the appendixes provide a methodology to achieve this higher level.

1.5 INDUSTRY PROFILES AND STRATEGIES

1.5.1 Automotive

New product strategies can take on many forms and philosophies. Among the various industries studied, the example of the automotive manufacturers tends to follow two different directions, that of full-line versus niche or limited-line marketing. Both domestic and overseas firms have advocates in each camp. Full-line manufacturers tend to provide something for every taste and want. Every size class of vehicle is covered, from small subcompact cars, to midsized sport utilities, to full-sized pickup truck and large sport utilities. The resource requirements to cover each market segment can be enormous, if all lines have unique platforms, that is, underbody frames in the case of body-on-frame vehicles or front, center, and rear floor pans and rails in the case of unitized body vehicles. Decreasing life cycles are also affecting manufacturers trying to offset new product development costs with sales revenues that barely reach maturity before the product becomes obsolete.

Some niche manufacturers tend toward lower volume or specialty vehicles that share components with other manufacturers' platforms, occasionally with common drive lines, or at least engines, but with unique exterior ornamentation, i.e., grilles, bumpers, headlamps, trim, and moldings on hood, fenders, quarter panels, and so forth. Others specialize in high-volume models that retain essentially the same design year after year, with very little change in either content, trim, or options. What defines them as a niche manufacturer is the fact that they have chosen not to produce models to compete in every segment, or that decision is dictated by resource availability.

The issue of brand management has to be considered in terms of the market. Worldwide consolidation of manufacturers has caused major upheavals among some established brand names, which are in danger of being squeezed out by the actions of the mergers, or by competitors by growing some models and shrinking others. This has contributed to the dilution of some brands, particularly those without a full line of vehicles to compensate for market changes; i.e., when one line is down, another may be up. Global market planning may also limit the scope or speed at which local-level adjustments or improvements can be implemented. Some manufacturers are struggling to support older, obsolete models with cosmetic "dressed-up" or minor redesigns, just to give dealers something new to sell, even though the market has already bypassed these vehicle platforms. A major philosophical change in a manufacturer's strategy is the only way to break out of this lockstep approach; often this requires a complete change in the executive ranks.[1]

Other factors that make new product strategies unique have to do with the difference in goals, resource utilization schemes, financial arrange-

---

[1] Juergen Schremp, "How, Why of DaimlerChrysler," *Automotive News*, October 26, 1998, p. 14.

ments, innovation skills, marketing expertise, etc. If the organization is set up to anticipate, discover or adapt new product ideas, then it should not suffer from a lack of new products. How the process is performed, what processes are created or changed, where they are done and by whom— all can play a significant part in the success or failure of the new product program.

On a global basis, while sales revenues fluctuate widely from year to year, profits remain low compared to other manufacturing industries. Some companies have had extremely low or negative profit growth for many years. This has resulted in the need for government or parent company subsidies to remain in business. A further result is a projected overcapacity of more than several times what is actually needed, worldwide, to replace vehicles lost by total-loss collision, theft, or scrapping. Some of the reasons for keeping marginal operations alive—in the face of mounting competition from newer, more efficient facilities located in lower-labor content areas—have more to do with political decisions than economic ones. An automobile industry has become a source of national pride and power in some countries, even if the operation is only a CKD, (i.e., completely knocked-down export model, a kit that needs to be assembled), because it also represents a large employment potential, including suppliers of parts, service, and infrastructure. Some manufacturers utilize single sources for basic components such as starting motors, alternators, and so on, which can be produced in standard sizes and capacities, at low enough costs to more than justify the added logistical costs of shipping and inventory, from halfway around the world. Export exchange-rate fluctuations can be a concern when local economic conditions worsen.

Another factor for a particular strategy has to do with integrated manufacturing. Several large conglomerates not only produce the basic vehicle shell in their own plants, but also the power train, electronic components, seats, and trim items. One of the potential offsetting elements of this strategy is response time to changes in the market. Quite often, outside suppliers can react much more quickly than the company's own internal parts operations, and with far better designs, lead times, and costs than the corporate offering, when new products are needed in a hurry. If design and engineering activities are also integrated on an international scale, response time, quality, and delivery schedules may have a significant impact on new product innovation plans.

Examples of this activity are illustrated by the number of suppliers involved in the development of modular construction techniques, where large subassemblies or modules are designed, manufactured, and assembled by the supplier.

Partnerships, marketing agreements, and joint manufacturing operations can provide additional, alternative avenues for new market penetration, without the large capital investment requirements necessary for factory ownership. "Badge engineering" is the term usually applied to this activity whereby sister vehicles share virtually all components except the nameplates or "badges" on the outer panels. Development expenses can still be considerable for a completely new product program, requiring 50,000-mile (projected soon to be 100,000-mile) power-train emission durability certification, new safety standard compliance, accelerated suspension durability testing, body corrosion, and so on. Anticipating future regulation and the continuing need for manufacturing cost reductions have provided a clear path for some original equipment, that is, vehicle, manufacturers (OEMs) in their quest for competitiveness.

Supply-chain integration is an example of how an OEM can gain competitive advantage. This concept seeks to consider all transactions between members of the "chain" of product development as one entity instead of many; these include funds, training, processes, information, and culture. The goal is to improve the collective performance of the chain without sacrificing or destabilizing any links in the chain.

There are several other factors that could have strategic consequences. These include upheavals in the distribution channel, depreciation of turned-in leased vehicles, excessive consumer debt loads, and service technician shortages. Some of the past concerns that have faded but might return include fuel availability, cost, insurance costs, and more stringent safety, emissions, and fuel economy regulations. Global warming is also part of this trend, as the world's leading nations try to arbitrate a treaty that is fair but rewarding to progressive economies.

Dealers are now faced with customers wanting to purchase vehicles, specified through Internet connections and ordered from hundreds of miles away but who may never bring the vehicle back for servicing. They may even purchase the vehicle over the Internet and expect the dealer to deliver it at virtually the dealer's wholesale price.[2] Turned-in lease cars can cause a temporary surplus in the market, which may depress prices of other comparable used cars; this in turn drives demand away from new cars, resulting in the need to discount them, thus lowering profits, and so forth. The main reason leasing has become an issue is the inability of consumers to finance new car purchases, which may be related in part to their current debt load. Leased vehicles can turn over more quickly and depreciate faster than pur-

----

[2]Ralph Kisiel, "Dealer Caught in Web—Gladly," *Automotive News*, Sept. 26, 1998, pp. 3, 63.

chased vehicles, resulting in a residual value that frequently exceeds the amount financed. This leads to a net loss that directly affects the profitability of the lessor. Interest rates can be another factor affecting the viability of leasing; that is, the fees generated in the turnover depend on the cost of capital.

Service technician shortages are due to two separate but related causes. The first is a shortage of people who want to work on cars, in general, due to their lack of interest in manual labor. The second factor is more critical; an inability to find or train enough technicians qualified to do the electronic system diagnosis required to repair modern vehicles. The education level of the average technician is currently insufficient to comprehend the system's operation and to troubleshoot warranty problems with it. The number of highly skilled technicians who do have what it takes is so small as to have caused a shortage, in just the dealerships, currently estimated to be in excess of 50,000 and rising.

Questions of crash compatibility between different-sized vehicles could be, perhaps, the most potentially devastating recent development to the industry. This is caused partially by the influx of off-road, four-wheel drive, truck-derived sport utility models involved in head-on and side impacts involving two vehicles. The first is typically a small car, and the second a truck with a higher weight and/or ground clearance advantage, which can produce significantly higher injury and death risks for the driver of the smaller car. However, statistically, these collisions are infrequent. Nonetheless, new regulations may be anticipated that force manufacturers to redesign larger models with devices that minimize the increased intrusion depth problem, particularly with side impacts.

In summary, the worldwide automotive industry faces severe challenges in the areas of financial instability, overcapacity, supply-chain consolidation, antiquated facilities, aging workers, shortening product life cycles, and looming environmental restrictions. Many of these issues cannot be overcome with faster and more efficient new product development, linked by Simultaneous Engineering. However, if a significant number of the vehicle design factors affected by the outside influences can be quantified, then there is an opportunity to provide the market with timely, creative solutions that may help solve the larger societal problems. The case study that follows in Chapter 11 pursues such a solution.

1.5.2 Aerospace

The aerospace industry, with a very different structure from automotive, is characterized by only a few large, well-integrated manufacturers. Both civilian and military aircraft production have evolved the same way, since most competitors have been involved in both ends of the business. From approximately three dozen major airframe manufacturers, incorporated in the first 50 years of the 20th century, less than a dozen remain, either as large independents or as part of other firms. The end of the cold war has resulted in fewer defense related contracts worldwide. In some cases, the contracts are substantial, designed to protect and nurture the net worth of the receiving company for many years. The other factor has been the on-going merger and consolidation activity, for the purposes of improving stockholder equity, increasing revenues, and reducing production and marketing costs. This has occurred not only in aerospace (including suppliers of airframe components, jet and piston engines, electronic systems, seats, interiors, and so forth) but also in nearly every other major industry.

Deregulation of the airline industry has also produced several dramatic side effects for manufacturers. Besides the inevitable increase in competitive routes, the demand for newer and more planes has caused a backlog at the manufacturing plants. With only a small number of aircraft builders, lead times for delivery of large orders are spread out over several years. Depending on the economic conditions of the local market and the financial position of the company, orders can be canceled, delayed, or shifted to other products, which often results in penalty clauses, late fees, and other costs. The airframe manufacturers and airlines negotiate these changes on an ongoing basis. The wide variability in new airplanes ordered can also cause massive human resource upheavals as the manufacturers try to match worker employment to the orders on hand.

All airliners are built as "custom or semi-standard," which includes limited variations in overall dimensions and customer-designated seating capacity, interior appointments, navigation system options, service options, cargo capacity, and so forth. While the basic package appears to be almost modular, there is sufficient design and engineering work required that each new derivative is considered an all new product. Engine suppliers, in particular, have capacity, specification, and subcontractor issues that are often negotiated, in concert, with the airlines and airframe manufacturers, who each have their own engineering staffs.

Opportunities for practicing simultaneous engineering abound in this atmosphere. Basic specifications and capabilities need to be tuned to the needs and wants of the market. Commercial airliners may appear to some people as "buses with wings." However, the safety and crashworthiness of airplanes, always paramount in the eyes of the manufacturers and the airlines, have become noticed by the public at large as the result of a major airline crash, somewhere in the world, virtually every few days or weeks. Small airplane manufacturers are not immune to the same media exposure. Lack of regulation in maintenance activities and pilot errors and training deficiencies tend to be the primary causes, but the airframe manufacturers are caught in the web of "deep pocket" litigation practiced in North America, in particular, and now spreading to the rest of the world. Crash survivability has become an important design parameter for airframe manufacturers, prior to any federal regulatory activity. The trade-offs with other parameters sensitive to the operational "envelope"—altitude, fuel consumption, range, and payload—are critical to the economic value of the product. Any significant increases in weight to improve safety may need be compensated with corresponding weight reductions in other areas; otherwise, airplane performance and customer satisfaction may suffer, particularly if other airframe manufacturers can find a solution involving a less restrictive trade-off.

Other issues confronting the airframe manufacturers and established airlines are the proliferation of new or spin-off carriers; local, regional, and national economic conditions; leasing and buyback credits of new and old planes; lack of progress in future suburban and rural airport facilities development; antiquated takeoff and landing control equipment and controller training. All of these issues can and do adversely affect the strategic plans of both the airlines and airplane manufacturers. These items can also positively affect plans if senior management can build in a measure of flexibility, to allow for sudden and constant change. Chapter 12 gives a fictional account of how an aerospace company develops a new flexible airliner design that is adaptable to several regional and business jet applications.

1.5.3 Military/Heavy/Off-Road Vehicles

This industrial segment is actually several subsegments, classified together based on the types of vehicles developed, customer groups, usage patterns, marketing plans, and so on. Vehicle characteristics and categories are listed below:

Load carrying
Drive system
Power-train type
Intended usage
Special equipment

Load carrying could include passenger, cargo, specific, or a combination of all three. Wheeled vehicles include buses that are primarily designed for carrying passengers but can also carry express freight packages and other small, light cargo. Over-the-road trucks can have cargo containers mounted behind the cab or on a separate and multiple trailers. Heavy cargo trailers are either flatbed or depressed-bed for transporting large, heavy objects.

Specifically designed vehicles have unique bodies adapted for special equipment, such as cranes, dump boxes, recovery winches, liquid handling, recyclable materials, fire equipment, rescue ambulance, military applications, and so forth. There may be common features among these various vehicle types, including power train, front and side exterior panels, drive systems (i.e., two-, four-, six-, or all-wheel/all-track drive) load frames. Whether privately owned or corporate or government, some may be for personal use, such as motor coach recreational vehicles, but most are used for commercial purposes.

The key factor distinguishing heavy-duty vehicles from light-duty is a major contribution of one or more customers in nearly all phases of their development. While "standard" products are still a mainstay in most companies' inventories, this category has become more of an exercise in semantics. If a major customer has strong convictions concerning the potential redesign of a "standard" product, it may not be long before the company gets its new product development team into action.

However, totally new products in the heavy vehicle market segments are not often introduced with the same frequency and flair as consumer products. One reason is the development cost. Another is the nature of the heavy vehicle market, where these products are typically classified as capital goods purchased as business assets to be amortized over their useful life, usually 10–20 years. Actual life cycles are typically much longer, since it is not uncommon for the original manufacturer or other more specialized firms to rebuild, retrofit, or refurbish units still in service. These actions involve power trains, drive systems, hydraulic or other auxiliary power units, electrical/electronic components, body panels, interiors, instrument displays, and other operating control devices.

Based on the situation in the heavy vehicle market segments, the main issue concerning new product development is not rapid, radical change but more orderly progressive introduction of extensions in existing product lines, new features to accommodate influential customers and provide enhancements in service, warranty coverage, and other after sales support activities. Also, older products are supported long after light-duty models would have been completely replaced.

Opportunities for Simultaneous Engineering are also prevalent in this industry. It should be driven and strongly represented by the service and warranty sectors, so that potential new improvements do not detract from basic operational capabilities of the equipment. Chapter 13 describes a case for a universally adaptable tracked vehicle.

1.5.4   Electronics

The electronics industry is multidimensional, multimarket, multiproduct and quasi-customer-friendly. Electronics are so pervasive and necessary that virtually every product either contains an electronic module or interfaces with one. Similar to the preceding industries, the electronics industry began around the turn of the current century, with the discovery of vacuum tubes that could be configured as diodes (i.e., a one-way valve) a triode, an amplifier, and many other applications including receivers, filters, inverters, and so on. These components were initially assembled into audio receiving sets replacing crystals, conversion devices and later into video. Proliferation into recording, transmitting, photographic, and duplicating devices occurred, driven, in part, by military requirements. Miniaturization was also necessary, due to the need to provide air transportation of advanced weaponry, larger payloads, higher altitudes, and greater range. Space exploration accelerated the process, with the discovery, after World War II, of some of the more useful properties of semiconductors such as germanium and silicon. Small quantities of these materials, plated into junctions between metals, can produce the same effects as the vacuum tubes, listed above, but at a fraction of the power, size, and speed of the tubes. Later, it was found that complete circuits could be combined on the same piece of silicon "chip." These devices, called "integrated circuits" (ICs), have been responsible for even further consolidation and combination of functions, including computers.

ICs are the central component to be analyzed. The manufacturing of silicon-based circuits is the central theme of this analysis. Other components in the typical electronic system (output, display, storage, input, transfer, and power) are considered to be performing functions that are peripheral to the IC. The main functions of ICs are the following:

Switching
Activation
Amplification
Tabulation

Storage

Calculation

Transmission

Control

Progress in the development of advanced manufacturing technology for ICs has led to production of large-scale ICs with literally thousands of the above functions available on the same physical surface, or chip. This development has, in turn, led to the utilization of very large-scale integration (VLSIs) in many new devices including personal computers, HDTV (high-definition television), compact disc-read-only memory CD-ROM drives, cellular phones, and so forth. The size of these devices keeps shrinking as manufacturers find ways of compressing more functions and storage capacity onto smaller and smaller chips; this has caused the cost of manufacturing to decrease along with the size. An ancillary result of this size reduction and capacity increase is the need to insure that the new processes do not cause deviations in quality and reliability. Another issue, related to marketing, is the time lag to get new products to the market during a time of high innovation; that is, the new product could become obsolete even before it reaches the market. A vast infrastructure is required in order to support IC manufacturing. Overseas suppliers are often relied upon for small components, hardware, and assembly labor, due to a favorable exchange rate and labor situation. The "pipeline" can be sometimes slow to react to market changes, which can result in large quantities of unsold merchandise, heavily discounted or liquidated, if bankruptcy is the only survival alternative.

The present situation is a debate concerning whether technology is driving the market or the opposite. The issue is somewhat confused due to the perception that there is a real customer demand for new products with more computing power, higher-resolution video, wider-range sound systems, more features and smaller packages than the previous product offerings. A point of diminishing returns can very quickly be reached if a product's life cycle becomes foreshortened sufficiently to preclude generating enough sales revenue to pay for the direct manufacturing costs, let alone development and marketing expenses.

There are several other long-term issues affecting IC innovation strategies. One is software, in particular, the digitally coded operating system for computers and virtually all other electronic product operating programs. Chapter 14 explores these issues in a case study in which a PC is designed to fit into different transport vehicles.

1.5.5   Retail and Packaged Goods Products

These industries consist of literally thousands of large and small firms involved in the manufacture of a wide variety of products for consumers, offered for sale in supermarkets, hardware stores, specialty shops, mail-order, and now, over the Internet. Since the diversity of items is sufficiently extensive as to defy analysis, some restrictions have to be put in place.

First, concerning food items, only preserved, packaged products will be considered. These include cereals, snacks, coffee, tea, beverages in cans or bottles, single and in multiple packaged quantities thereof, sauces, soups, baked goods, baking supplies, condiments, spices, dairy-case meats, milk, ice cream, cheese, and other dairy products, and so on. The second group includes paper and plastic products, cleaning supplies, medicine sold over the counter, cosmetics, hygiene products, hardware, small appliances, toys, and tools. All packages have to be sealed, dated, branded and/or serially identifiable with bar codes.

The rate at which these items are developed, introduced to the market, and tested (and in many cases, withdrawn after a few months) would indicate that a major shortfall is adequate and accurate market research. Aside from the problem and expense of gathering and analyzing existing market data, another factor has to be the allocation of shelf space in the stores. The competition can be fierce, and the major brands neither ask for nor give any quarter to their competitors. Discounting (both wholesale and retail), sales, and redeemable coupons are some of the main competitive actions that are a part of every day life in the aisles of the supermarkets.

A new feature that is just becoming valuable to both the retailers and manufacturers is the database derived from the bar codes, read at every purchase. The raw information has been available for several years, but only recently have firms offered analysis of the data as a service. Besides recording real-time purchase patterns, tracking of new product introductions is feasible, so that manufacturers can gauge very quickly whether the latest new product is meeting its objectives. Other key issues affecting retail products are safety, convenience, turnover, newness, and profitability. Chapter 15 provides specific details, concerns, problems, and solutions for packaged consumer goods issues.

The appendix contains a checklist of simultaneous engineering items pertinent to each phase element of the new product development process, beginning with strategic planning, market research, idea generation, and so forth.

The next chapter will explore how manufacturers identify their markets and key opportunities for new products and for Simultaneous Engineering to facilitate their development.

The rest of the current chapter is the beginning of a mini–case study of a fictional company called Ajax Manufacturing. After each of the first 10 chapters, the last section will provide an update on how Ajax is contending with the problems and opportunities presented here.

1.6   AJAX MANUFACTURING COMPANY

Ajax Manufacturing Company was founded by a university professor in the mid-1930s as an adjunct to his research on surface roughness and the interactions with tribology, the study of wear on machined metal surfaces utilized for bearings or sealing. The company grew exponentially during the World War II and cold war eras, as the defense industries geared up to perfect better and better weapons systems. Later, faced with new competition from import vehicle manufacturers, the domestic automobile manufacturers began to explore new quality and reliability initiatives, involving new manufacturing methods to control or eliminate out-of-tolerance components. Measurements for surface roughness, waviness, and roundness were all deemed necessary to control warranty expenses on failed engines, transmissions, and so forth. Ajax built several lines of instruments designed to provide these measurements on virtually any machined surface. Even after the original owner retired, the business flourished for more than 40 years, with only occasional lapses coincident with business cycles and competitive inroads. However, within the last 15 years, Ajax lost significant market share to more aggressive marketing efforts and instruments of superior quality of both design and execution. After several years of watching their market disappear, the company was sold to a group of investment bankers, who believed that Ajax still possessed the means to compete on a global basis, albeit with a narrow, niche product. The senior management team assembled a group of current and former employees, major customer representatives, and marketing and technical consultants to develop a new strategy for the company. The group was referred to as "the new product development committee."

The new product development committee was formed under the leadership of the marketing vice president. The first meeting of the strategic planning subcommittee, chaired by the president, took place about a week after the initial gathering. Besides various department managers, selected customers, suppliers, and representatives from Ajax's investment banking firm, their accountant and several consultants were invited. After introductions, the chairman discussed the following agenda:

Overview of current and future global economic, industry, market, and product situations

What is Ajax's vision? What should it be?

How should go about changing our strategy?

What additional resources will be required to accomplish the new strategy?

How long will it take to implement the new strategy?

The first presentation was by the chief economist of the company's parent bank from an international cartel. He discussed the current global economic situation, the effects of large-scale mergers and acquisitions, the Eurodollar implications, future energy-cost predictions and other concerns, raised in the question period following his talk. He stressed that, while consumer confidence seemed to be at an all-time high, a crisis could be coming soon in some capital-intensive industries. Overconfidence was tending to fuel an overinvestment in new plants and equipment, not only in North America but also in other markets known for low labor costs. However, the monetary crisis in Asia was countering this enthusiasm with widespread chaotic conditions, deflating currencies, financial market panics, and workers either being laid off or having wages cut by more than 50%. Aerospace, automotive, building materials, chemicals, electronics, paper, textiles, toys, among other industries, were all in a disarray due to different combination of the above factors. The global nature of these fast-paced developments meant that the crisis would no longer be confined to a given region but would, in time, spread everywhere. The trick was to be able to accurately predict when and where the crisis would hit next. For many Asian businesses, this prediction would decide whether they stayed in business or not.

On the domestic side, while consumer confidence remained high, buoyed on the strength of the economy and with unemployment at historically low levels, spokesmen for some those same industries were acting as though a recession were not only coming but already here. The chief economist described some of the most recent megamergers between industrial giants as catastrophic for redundant workforces. This was in spite of reassuring words from executives trying to dispel persistent rumors of major layoffs. However, this gloomy outlook was not shared by all larger firms, since many were pursuing extremely aggressive expansion plans, not only in Asia but also in North America. New manufacturing plants, expanded capacity, and renovations were prevalent across the North American continent. Recent conferences were the subject of this expansion, which did not always seem to have a completely economically based justification to it. Decisions were being made, he intimated, as much based on a euphoric belief in the future as on hard data. "It can't happen to me!" summed up the general feeling about any future downturn.

The bank economist concluded his remarks with a stern warning that at least his institution would not be as liberal as it had been on any overly optimistic expansion plans, but that he would council with Ajax executives and help them decide the best course of action.

Next, a consultant discussed both general and specific trends in the scientific and industrial instrument markets, competitive product movements, and overseas trends within the last 18 months and predictions for the upcoming 12 months. She went through an electronic slide presentation that showed major market shifts, with both Asian and European instrument makers concentrating new product introductions in North America. This was due, in part, not only to the Asian market slump but also to a more aggressive effort by the Europeans, whose economy was just recovering from a long recession. The cause of the shift was also shared by an awakening, by domestic manufacturers within the last 15 years, of the need to improve the quality of their products. In a global economy, where domestic products typically suffered from cost and price disadvantages, any product defects, excessive warranty claims, or other quality problems were unacceptable. This attitude had now pervaded the domestic market, where exposure to higher-quality Asian and European goods was upsetting the situation and causing local plant closings and layoffs. Most of these were either short-term, as the companies learned to do business in a new light rather quickly or left the market permanently. The remaining players were marketing new products at a remarkable rate, by taking advantage of the new "quality initiative" being introduced by the transportation industry, in particular. Fabrication and assembly-line personnel were being given more responsibility to decide whether products were "acceptable." This process entailed their ability to "stop the line" if they saw something wrong.

Thus, with all such quality inspectors around, the need to increase measurement capabilities became a natural occurrence, along with required training, which was requested by the unions. This, in turn, gave rise to an increase in orders for new instruments, mostly handheld, for measuring profile or shape, roughness, waviness, and so forth, of machined components. Previous instrument utilization was mainly in random, spot checks on semifinished parts, on workbenches or in metrology laboratories, and only by quality control department personnel. This was partly due to the ungainly orientation that the work piece had to assume before measurements could be taken, since they generally had to be removed from the lathe or milling machine prior to measuring, then returned to the machine, if more material needed to be removed. Also, the instruments were not easy to use, and were slow and unreliable.

The major 35–40% increase in these orders was causing potential suppliers to concentrate on sales efforts and marketing programs. The result was a 10–20 fold increase in the sales of these instruments to the aerospace, automotive, electronics, rail locomotive, heavy truck, and construction machinery market segments. She explained that the market showed no slack but, on the contrary, seemed to show a increasing level in the overall demand curve.

The other presentations by customers, suppliers, and employee representatives weaved a consistent story of haphazard quality; difficult relationships with suppliers, customers, and other companies; and contentious negotiations with workers, both union and salaried. Ajax's reputation in the investment community was characterized as being "out of date, out of touch, and backward." The only hope, according to a consensus of the committee, was either a complete reorganization or acquisition by another firm. This did not come as a total surprise to the management team. A debate then ensued about the only apparent decision to be made: whether to wait until an acquisition became inevitable, or reorganize now and better the financial and industry position of the company.

The president interjected his thoughts to squelch this argument. He told the committee that he would not, in any way imaginable, allow the firm to degenerate further and would, furthermore, be willing to reorganize it personally, if that were possible, or dissolve it immediately. He then asked for a show of support and received it unanimously. He further proposed that a new vision for Ajax should consist of a quest to be the best instrument manufacturer in North America, on the basis of producing the highest-quality, best, and easiest-to-use products and of having superior relationships with all constituents. All in the committee pledged their support and energy toward these goals.

While the reorganization was being planned, it was decided that the new product committee would oversee the entire development cycle of the new instrument, designated "M-2000."

The next phase in the new product development process is market research, which will be reviewed at the end of the next chapter.

2

MARKET RESEARCH

2.1 INTRODUCTION: WHAT IS MARKET RESEARCH, AND WHY IS IT NEEDED?

Market research can be defined as the various methods of providing as much information as required by marketing management to make effective decisions on new products. Market research is mainly concerned with providing answers to several major questions:

What are the customer's present and future needs?

How will future needs be met?

When will they be met?

What are the competitors' current and future product positions, and how will they change?

Other indirect but related issues have to do with how, when, and where the various questions are asked. These issues can be as important as the content of the questions. Data analysis and interpretation are also significant, since a misinterpretation could lead to erroneous conclusions and expensive new product decisions. The timing, cost, and accuracy of research data make these decisions critical to the future profitability and growth potential of the firm. Answering all of these questions successfully usually requires the development of a systematic plan that can be administered and completed in a timely manner with meaningful results. All of this is predicated on the quality of the plan and the diligence with which it is carried out. This first step is defining the research requirements.

2.2   DEFINING THE RESEARCH REQUIREMENTS

The main issue with market research is to ensure that the most pertinent information is gathered on a timely basis, without having to sort through or pay for information that is useless, either because it was gathered along with the useful data or because mistakes were made in the gathering process. The research plan needs to be developed prior to any surveys, to eliminate as much confusion over research goals and objectives. The plan should cover the following steps:

Define the problem.

Determine the objective(s) of the research.

Develop the method of collecting the data or surveys.

Determine the sample size or population requirement.

Conduct or monitor the data search or survey.

Analyze the data.

Analyze the results.

Depending on the scope of the problem at hand, the necessary information may be obtained through internal company records, competitive intelligence, or research. Sales information on existing products may also contain facts that would pertain to a new product offering, particularly if the new product was to be a revision or minor redesign of the current product. It is more likely the case that a substantially new product is deemed necessary to meet future market needs and wants. This level of information definitely requires a more in-depth analysis of data, acquired through special efforts such as competitive intelligence or formal market research.

It is important that the marketing management think through every information request very carefully so as to define the problem as precisely as possible and to control or eliminate any waste in the information gathering phase. Reaching a consensus on the objectives of the research can be the single most difficult part of the process. This is due to the unpredictability of how the data will be analyzed and presented. In spite of having a plan, rigidly adhered to, research data frequently gets into the wrong hands or is more or less mistakenly or deliberately misinterpreted.

For example, poor sales performance for a recently introduced new product could be the result of one or more factors, including but not limited to the following:

There was inadequate product distribution at introduction time due to shipping problems.

Misleading or incorrect media information was distributed to the public.

The product launch was delayed by manufacturing problems related to late design or engineering changes.

Profitability suffered because product development costs greatly exceeded the original estimate.

The product does not meet functional design performance or regulatory requirements.

Dealers or distributors were not properly informed on product features or performance.

Warranty campaigns and defect recalls reduced public confidence in the product.

Competitive new product offerings have absorbed the target market share of the new product.

Each group involved with the new product could be looking for research data that would shift the responsibility for the poor launch to some other party, when in actuality, all parties probably share in the result. Open communication between the groups hopefully limits this kind of wasteful and pointless exercise.

Market research is not only done prior to the start of a new product program, but performed continuously, at each stage of the development process. Central to the Simultaneous Engineering philosophy is the continuous update of design decision information as each new product development phase element is thought through or completed. As the product concept is more precisely defined, additional questions should be presented, first to the internal design staff, and then to other in-house contributors, selected suppliers and customers, along with other similar outside groups that were utilized to critique the concept that formed the basis for the original idea. While it may not be feasible or practical to query the exact same outside people as were utilized originally, they should satisfy the same general selection criteria as the original group, within reason. Similarly, tracking the effectiveness of a prototype or beta market test, the impact of media prior to product launch, and competitor reactions are all topics potentially worthy of research.

Formal market surveys are not the only method that can be used to investigate the above points. Competitive intelligence is a method of obtaining information on competitive products. In the industrial sphere, competitive products are purchased, disassembled, and studied for potential advantages for a new product. Manufacturing cost, fabrication details and assembly complexity, repair, service, and warranty issues are all compared to the new product concept for any future design improvements. If competitive products have not yet been introduced, other means of analysis have to be pursued. Sometimes, suppliers of common parts may have knowledge of new competitive programs. Likewise, employees of the competitive firm may be willing to voluntarily reveal internal secrets on new product programs during job interviews, interindustry meetings, or other gatherings. *However, the practice of requesting confidential information should not be permitted and should be considered unethical in nature and not to be encouraged or tolerated by associates or middle or senior management.*

Beyond competitive intelligence, formal market research is the usual method for determining customer wants and needs, the acceptance of new ideas, and to whether a new product fits into an existing market. The need for the information has to be balanced against the cost of obtaining it. Formalized market research, as conducted through major information service providers, can be very expensive if the costs of data analysis and interpretation are included. Sample sizes from several hundred to several thousand final participants are typical. If the proposal from the provider is based on the total number of participants to be queried, the costs can be even higher. However, many companies have in-house research staffs that can provide their own analysis, as well as survey design.

Another issue involving Simultaneous Engineering is the anticipation of design-related questions that may need to be answered during the new product development process, prior to a final review and sign-off. Refinement to an initial product design could, under ordinary circumstances, take several years and result in many iterations before the final debut. However, skillful application of research could eliminate the intermediate, wasted motions and get to the final configuration and specifications sooner than later. The key is to define the problem at hand precisely and to get the right questions written down properly and queried in the right atmosphere.

Obtaining agreement from all partners involved on the expected outcomes of the research is important. This milestone has special significance from the viewpoint of Simultaneous Engineering, because all of the issues that could affect new product design, involving customer input, need to be addressed here. These issues need to be included in the following section on developing the research plan.

As will be seen later, some inquiries are received in the form of open-ended questions, which will require rewriting to control the data flow by eliminating ambiguities. Survey questionnaire design issues, which will be covered later, include open-ended questions, which seek a respondent's opinion, or questions with vague terminology, where the potential for misinterpretation is considerable.

2.3   DEVELOPMENT OF A RESEARCH PLAN

Market research project management begins with the development of the research plan. This multistep process lists all major tasks that need to be completed and the optimum order in which to accomplish them, typically, but not entirely, sequentially. Actually, the research plan should resemble a new product life-cycle process. The introductory stage of a new product is essentially when the planning phase of the research should be performed. Concept and product testing and name, advertising, and package validation are some of the early tasks that need to be performed before final product design decisions are made. During the growth stage, adjustments to the production schedule, refinements to the product, and redesigns are some of the possible outcomes of research. The maturity stage of the product may require research to determine if any potential exists for renewing the line through extensions, segmentation changes, new advertising, packaging, and so forth. Finally, the decline stage is where research can benefit the efforts toward salvaging the product rather than developing a completely new product. It can be in essence, the first stage of the new product design phase, closing the full circle in the new product development process.[1]

Planning the entire new product development process should involve all parties potentially affected by any market research proposed. The marketing department needs to provide the coordination of the plan, to assure continuity as well as a focus on the intended outcomes. This step, however, does not usually come first, but rather after the initial problem definition phase. Typically, the Simultaneous Engineering opportunities identified are related to the following areas:

New product introduction

Product improvement

Development of additional products

Product repositioning

Targeting a different segment

---

[1] Jeffery L. Pope, *Practical Market Research* (American Management Association, 1601 Broadway, New York, 1988), pp. 13–15.

Redesigned packaging

Revised advertising

Research can be utilized to resolve the above issues as well as others, through group interviews and through market segmentation and product positioning studies. The plan then evolves into testing certain aspects of the new product:

Effectiveness of improvements over the existing product

Packaging effectiveness

Advertising effectiveness for retention

After the product has been introduced to the market, several measurements of its performance are recommended, in case changes need to be proposed. These include the following:

Customer awareness of the new product

Attitude toward the new product

Intent to make, and making trial purchases

Satisfaction with the product

Repeat-purchase intent

Advertising recall

Problem definition may be the most important research action, since all subsequent activities depend on its accuracy.[2] If the perception of a product's brand image is different between the company and its customers, any research conducted may be prejudicial and therefore inconclusive or incorrect. The firm must be willing to provide the means of upgrading and validating its own product data model so that changes can be accounted for correct and future actions planned on those results.

An example of a generalized market research problem might occur when an existing product line is purchased by another company. Any previous research information might be no longer available or be out-of-date, which means that new research would be needed. The particular product line might have applications in several different industries, requiring

---

[2]Philip Kotler and Gary Armstrong, *Principles of Marketing*, 7th ed. (Prentice Hall, Upper Saddle River, N.J., 1996), p. 116.

multiple survey forms, each one slightly modified for each industry. With input from the brand manager from each key market, the survey questions are formulated on the basis of internal company sales records, checked for the proper "statistical content," that is, closed-end questions that narrowly define the intent of the questions and the expected answers. Generally, the simpler the questions, the higher the rate of survey returns.

Ranking and rating questions are frequently utilized to quickly determine the differences between respondent classifications. Areas to avoid are highly technical subjects and those requiring a respondent's opinion or a conclusion, unless the information requested is vital for product redesign. Typical survey questions should relate directly to product attributes, features, operational characteristics, problems, deficiencies, comparisons to competitors, potential improvements, convenience, price, repair, warranty, power requirements, and any other issues related to the product; its design, manufacture, sale, distribution, or service.

The implementation phase involves checking the accuracy and coverage aspects by monitoring the administration of the survey. Random or spot checks should be sufficient for this purpose. The objective is to eliminate as many errors as possible in data collection, including problems with respondents, misinterpretation of answers, and surveyor shortcuts, that is, not asking certain questions or making assumptions or mistakes in obtaining answers.

Analysis of the data would be the next step, concentrating on the priorities established in the research plan. Questions calling for quantitative answers should be addressed first. Qualitative answers, which express the respondents' opinions, wants, and needs, should be handled next. If a mathematical model is to be utilized, the researcher must verify that each term or constituent is applicable to the particular case being analyzed.

Interpretation of the data is the most important phase of market research. The survey could be the most successful in history but still be doomed to failure if the wrong conclusions are drawn from it. Consensus between the various recipients or end users needs to be established early in the program, to avoid conflicts when unexpected data comes in. For example, the marketing manager may interpret the answer to a question on product price as the main reason for lowering it to remain competitive on future models, while the engineering manager may see it as the need to improve quality without increasing the price on the next model. Both interpretations could be right or wrong depending on the answers to other questions, specifically related to the other issues raised. In other words, there should be sufficient cross-references in other questions to act as confirmations of the trends observed. The research manager must be sensitive to these various, possible interpretations and make sure that the questions are asked in the proper manner to provide the information desired, even if further questions are needed as supplements to the original questions.[3]

Feedback on the results is the final step, enabling the surveyor to evaluate the effectiveness of the survey for implementing project improvement.[4] Confirmation of expected outcomes, discussion of disputed or unexpected outcomes, and plans for future research surveys should be part of this process. Finally, presentation of research results and conclusions to senior management provide additional support for the new product program, along with any new tasks that have been identified. New items could be additional derivatives or variations of the new model products, additional or different features, or something completely different from that intimated in the survey. New survey approaches such as changing the survey format or method may be necessary depending on the initial results. The intent, layout and direction of questions may need to be changed if problems are encountered with respondent reception or reactions to the survey. Ethics and religious and ethnic sensitivities all need to be addressed in survey reviews.

2.4   RESEARCH METHODOLOGIES

Research methods include focus groups, telephone and personal survey questionaires, trade shows, and other means. The effectiveness of any research information is based on three main factors: cost, timing, and content. All three have to be planned carefully and balanced to maximize the value of the research. If any one of the three factors is not planned or is out of balance, the research may prove to be ineffective. Each method has advantages and disadvantages. Surveys can vary from providing low-cost, good-quality data but with a poor flexibility and response rate (i.e., mail surveys), to having moderate cost, excellent control, and a good flexibility and response rate (i.e., telephone surveys). Personal surveys are also utilized; however, the interviewer sometimes influences the answers, and the cost can be higher than telephone and mail surveys. Focus group interviews offer advantages of more specialized research into respondents' thoughts and feelings, but can be problematic because of bias or intimidation by facilitators with strong personalities. Computerized surveys are

---

[3]Kotler and Armstrong, pp. 129, 130.
[4]Pope, p. 18.

being used for standardized information on closed-question formats.[5] The main point is to match the survey method to the research plan.

As an illustration of the sampling dilemma, consider the situation faced by the U.S. Census Bureau. Once every decade, the bureau is mandated to count the population. They are provided a budget, a time limit, and a requirement for accuracy in the survey. While all reasonable efforts are made to count everyone in each category, that is, by location, number of dependents, citizenship, and so on, it has been estimated that up to 10% of any given category may be inaccurate or not counted. The counting methods include door-to-door canvassing, cross-checking against other databases, voter records, and so on. Because the results of the census can be influential in the distribution of political and economic benefits among the states, the pressure to collect all valid residents has become enormous. Therefore, proposals are presented to collect and validate key statistics utilizing a variety of high-tech sampling concepts. Among them are several very sophisticated, computer-based analysis software to determine the quality of the census estimates. The level of quality assurance is a function of the program effectiveness (i.e., thoroughness), which can be equated to computing time and cost. In the end, the decision should be based on the incremental costs to validate the last few percents of residents. If collection and validation costs are significantly higher than the percent of suspected missing residents (i.e., an additional 25–50% of the cost of the total program, to verify that the last 10% of the data is good), the incremental cost is probably not justifiable. If the incremental costs are more reasonable, the situation may require consideration of other factors, such as timing or data analysis complexity.

2.5   SURVEY QUESTIONNAIRE DESIGN

After all of the questions have been collected, the next decision is how to present the questions and to what group. The typical options include telephone and personal surveys, personal and group interviews, and data analyses from product sales histories. There are advantages and disadvantages to each format. The weighting of each category must be evaluated and decided upon prior to administering any survey. Otherwise, the results may prove to be inconclusive.

The advantages of telephone surveys are cost and data collection time. The disadvantages are limited responses due to a lack of interest or to

---

[5] Kotler and Armstrong, pp. 124, 125.

unwillingness to participate on the part of the respondent, limited opportunity to identify biased answers, and limited ability to describe key product features and functions. The advantages of group interviews are the potential synergy of the group interaction, closer correlation to actual product usage, and the opportunity to identify potential new features through the observation of role playing within the group. The disadvantages include the presence of shy people to give any answers (making some members afraid that their answer would be considered "wrong" by the others in the group or by the "moderator") or the potential dominance by highly verbal or forceful members of the group.

The advantages of written surveys include higher levels of accuracy and descriptive detail than telephone surveys. Disadvantages include low response rates and long collection times.

The advantages and disadvantages of database analyses are due to the limitations of the data itself. If the data is sufficiently complete, then a more detailed analysis is possible. However, having data seems to create the need for more data, which can cause delays in the completion of the project while collection is attempted. It is difficult to imagine having too much data, but it happens, with the results delayed to an unacceptable level. The key decision is knowing when enough data has been analyzed and the proper questions answered, so that the next step can be started.

2.6   INITIAL NEW PRODUCT RESEARCH

When preparing for a completely new product effort, existing databases are available for developing a basic research strategy. These databases consist of computerized listings of product and market categories, arranged into standard industry classifications (SIC, now known as NAIC, or North American Industrial Code System), codes. Regional and geographic data categories are also available at public libraries, universities, and now electronically over the Internet. Likewise, information on competitors or potential competitors can be found in trade association journals and reports funded privately by specific industries.

2.7   STATISTICAL SAMPLING PLANS

When data has to be retrieved from customers or potential customers, a sampling plan needs to be developed. If the total population (i.e., the number of recipients), is small, direct contact by phone or mail is probably the most cost-effective approach. However, in most cases, the population that needs to be reached is either unknown or too large to economically survey directly. In either case, a statistical sample must be devised, tested, and utilized for new product decisions. Sample recipients, households, or firms must be representative of the whole population. Sample sizes from a few hundred to a thousand have been relied upon to predict buyer and customer behavior toward new products. The main point is to minimize the effects of errors that can occur in the actual sampling process. Making sure that every sample chosen has accurate information related to the demographics and other key measurements is an important step.

2.8   ESTIMATING SAMPLING ERRORS

Sampling plans need to address two main issues: selection of the actual entities to be included in the sample, and extrapolating the results of the sample to the whole population under study. Estimating the errors from the various sampling plan also need to be addressed, both from the sampling itself as well as nonsampling errors.[6] Sampling involves selecting a group of data points or potential data transaction points for research. These data points can be chosen either at random or as the result of an ordered system. To ensure that a completely random or probability sample is selected, all elements in the total population under study would need to be known. Since this phenomenon is virtually nonexistent, a compromise is usually arranged whereby some measure of nonprobability is substituted. Thus, classification often results from screening survey respondents utilizing the following criteria for possible grouping:

Previous or current product ownership, or industry
or market experience
Competitive product ownership or usage
Geographic location
Demographic profile

Probability sampling would be the ultimate accuracy, if the entire population of a set of characteristics were known. However, since it is usually not cost-effective to uncover all units in the population, some error will be

---

[6]Pope, pp. 218, 219.

introduced into the sampling plan. The major decision required is to determine the sample size, based on the confidence limit desired and the cost to achieve it. As the sample size increases, the confidence range, or "error bandwidth," that the measured quantity is being more accurately predicted decreases compared to that from a smaller sample. This measurement typically reflects a mean or average quantity, with a bandwidth consisting of the confidence range (i.e., plus or minus the error calculated for a particular confidence level). The confidence level is selected to provide the accuracy desired for decision making, typically 95%, which means that the maximum scatter in the data is limited to two standard deviations. This level of accuracy is not too difficult to achieve, compared to a recent program to gain a six-sigma level of accuracy (i.e., six standard deviations), an almost unimaginable level of approximately 350 maximum defects per 1 billion operations! Yet, the manufacturing processes responsible for this milestone are in place for everyday household products.

An example illustrating concept of confidence range follows. Assume that a survey administered to 50 potential new vehicle buyers asked them to predict the retail price of a particular vehicle design configuration. The collected data showed that the mean or average retail price was estimated to be $21,675 \pm \$3,034$, or $\pm 14\%$. If 2,000 respondents were given the survey, the range would probably decrease to $\pm \$433.50$ ($\pm 2\%$), to achieve a 95% confidence interval. Obviously, it costs more to conduct a larger survey, justified by the need for greater accuracy. If the value of the decision warrants the expense, a larger sample should provide the additional margin of accuracy desired. However, it must also be noted that the law of diminishing returns applies. To collect an extra 10% of data may cost 50% more that the first 80% of data collected.

There are several precautions that address limiting nonsampling errors. The quality of the research is a reflection of the quality of the researcher and the people asking the questions and performing the analysis. The integrity of the respondents is another factor that can affect the outcome of the data. If they choose to not answer, give erroneous answers to some questions, or terminate the interview, the data is lost. Misunderstanding of questions, leading, or bias exhibited by the questioner could also influence the results, as could recording, coding, and editing mistakes. Most of these problems can be minimized through effective selection, training, and supervision of questioners and analysts. Clear, concise survey questions, pretested content, screened questioner applicants, and review of questions for Simultaneous Engineering opportunities will provide the best practices for achieving objective results.

2.9 DATA ANALYSIS METHODS

Data analysis consists of verifying that the variables selected for study provide the necessary information. It typically begins with cross-tabulating the results to find relationships between variables, such as relating a respondent's propensity to repurchase a product with the type of distribution channels in use, including department stores, discount stores, outlet stores, mail order, and so on. Demographic data could be compared between different buyer groups to find the mean or median age, income level, other product preferences, and so forth. Even technical questions can be utilized, if respondents are competent enough to provide the answers in a usable format. After the variables have been separated, a number of multivariate analyses can be performed. Selection of a technique depends somewhat on the distribution of the data. However, the distribution may not be apparent, which may require the application of several techniques in order to test the validity of the data. To assume that the data is normally distributed means that the average response in quite near the center of the total number of data points. There are several statistical tests that can be administered to confirm the hypothesis that a given mean represents the distribution it came from, within the confidence interval necessary to insure proper decision making, usually 95%. A 95% confidence interval means that data is significant if is concentrated within two standard deviations of the mean. One of the simplest tests is the student's t test, which has shown good correlation for relatively small sample sizes (3–32). It does not have the exact same shape in the tail regions as the standard normal distribution, but it is usually close enough for most applications that can utilize a small sample size.[7]

If the data is suspected to be not normally distributed, it may still be representative of certain trends, which could provide valuable research information. Hypothesis testing is one method for comparing data and for comparing data analysis techniques. Estimation techniques such as mean, median, (midpoint) variance, and standard deviation, are useful for both small and large samples. However, when the quality of the sample is in doubt, a hypothesis is utilized to test the "goodness" of the statistics of the sample. Various tests depict nonnormal data fields and specialize in determining particular abnormalities. The t, z, F, and chi-square tests differ in degrees of the abnormalities in their respective tails of their distributions

---

[7]William Mendenhall and Richard L. Scheaffer, *Mathematical Statistics with Applications* (Wadsworth Publishing, Belmont, Calif., 1973), pp. 283–287.

and how they are approximated. Basically, once collected, one or more tests are utilized on the same sample of data, to verify that a sample mean is representative of a larger but unavailable dataset due to sampling limitations. The development of these hypothesis tests and the application to specific data sets is beyond the interest of this text, but can be obtained from references in the footnotes.

Other tests have been developed to uncover the interdependence between variables. Typically, dependent and independent variables are both under analysis at the same time. Dependent variables are those describing phenomena such as product usage or customer satisfaction with a product. Independent variables are used to predict what will cause changes to the dependent variables. In some cases, there may not be a clear cause-and-effect relationship between the variables, particularly with a large number of them.

The most recognized test utilized for analysis of interdependence is regression. This technique seeks to establish a relationship between variables, whether it is linear, quadratic or some other mathematical function. Regression analysis does this by trying to fit a function to the data provided, by testing the linearity or proximity to a curved line or surface, utilizing the least-squares process. Other interdependence measurement techniques include factor and cluster analyses, multidimensional scaling, perceptual mapping, and conjoint analysis.

All of these methods offer ways to group responses by differentiation or segmentation into common or consistent elements.[8] These techniques offer alternatives to comparing and contrasting product preferences, where there are no clear-cut right or wrong answers, only opinions. Perceptual mapping extends this premise by allowing a multidimensional view of all the opposite choices in a survey result. If, for example, respondents rank both vehicle size and utility as significant buying influences, then the combination of large, high-utility vehicles might experience more of a clustering effect than a small high-utility model. A two-pair comparison is called a conjoint analysis, but larger numbers of opposite pairs can also be considered. However, visualizing a map of more than three dimensions can be complicated and potentially misleading.

In the spirit of Simultaneous Engineering, considering all choices together should be the obvious direction, if the analysis path is the same. If product preferences range across a multitude of choices, the strongest reasons for the preferences can be easily selected for further study. Frequently, the data is so plentiful that primary and secondary analyses are necessary.

---

[8]Pope, pp. 236, 237.

A primary sort might delineate several pairs of preferences, which would then be further broken down, in field tests, to verify that the analysis system is working as designed. Color and trim set combinations for cars would follow such a scenario.

After an analysis of the collected data, the final step is to present it to senior management for approval of plans to study either product improvements, product abandonments, or a completely new design. The key is to make sure the data reflects the intended outcome. Even if the data is nothing more than an analysis of past sales figures for the company's products as well as those of competitors, the presentation should be made in the most professional manner possible. Prepare an agenda, give a brief introduction, and summarize the main points, but with backup data analysis if required. After the discussion and question-and-answer periods, concluding comments should strive for a decision. Presentation media should be attention gathering without being overly flashy or too detailed. Some of the newer software packages have excellent subroutines for doing such graphics as histograms, pie charts, fish-bone diagrams, and perceptual maps, in full color.

2.10   AJAX MANUFACTURING COMPANY

The second meeting of the new product committee was held one week after the first meeting. All of the same participants as before were there, and in addition, more research staff. The agenda consisted of the development of a research plan for surveying customers of the E/D-12 and competitors for information on how what they liked and disliked about each instrument. Plan called for obtaining 100 responses to just 10 questions, designed to collect the maximum number of facts for a minimum of cost. All customers were to be called by members of the committee within three work days, if possible, to get the data logged in and analyzed quickly. They began with 150 longstanding, major customers, hoping to get 100 valid samples for analysis.

Summarizing the results of the survey, the committee discussed the following problem areas on the E/D-12, compared to the competition, prioritized by importance:

An inability to reach the surface while the work piece is still mounted to the machine tool (82.7%).

Measurements take too long to conduct (74.3%).

Measurements are not stable, fluctuate excessively (69.4%).

The instrument is too heavy and awkward to use on machine (56.8%).

Multiple passes are required to get consistent readings (48.1%).

The instrument breaks down frequently (41.9%).

There is excessive power consumption (37.6%).

Directional changes of the probe cause voltage spikes, affecting readout devices (33.8%).

Frequent probe-tip changes are required to find acceptable measurement range (29.2%).

Measurement scales are difficult to read; too many optional scales are available (16.5%).

The analysis consisted of discussing the above responses. While it was understood that most customers still required a device for bench measurements, the bulk of the daily production quality checks had reverted to utilizing a handheld instrument, on the work piece, in the machine. The next four items were all related to attempts at measuring on the machine, with the E/D-12. The breakdowns occurred most often with the motorized movement of the probe rod. It apparently had a tendency, after a few hours of operation, to either bind up or become loose enough to change the readings when the direction changed. The next two items were due to the size of the motor required to move the probe rod.

The final answer was the most difficult to analyze, since some of the same customers had suggested the additional scales so that the E/D-12 could take comparable measurements with competing instruments. The Ajax engineering staff did not always agree with the need for the additional measurement conversions, but they considered them for incorporation as directed by senior management, to keep key customers satisfied.

The next step for the new product committee was to decide on a course of action for improvement, by generating as many new ideas as possible, as quickly as possible.

The next chapter will discuss how to find new product ideas from the data provided on past sales, trend analysis, and the comments made in light of the design reviews and focus groups.

3

IDEA GENERATION

3.1 INTRODUCTION: HOW IDEA GENERATION IS RELATED TO MARKET RESEARCH

Idea generation is, in reality, a very specialized extension of the market research process. It has become a unique subset of research due to the nature of its intent; to generate specific new product ideas that can, later, be screened for all of the factors that determine whether or not the new product will succeed in the market. The two most important parts of this strategy depend on the proper selection of participants and controlling the flow of new ideas without allowing any evaluation by the participants during the collection phase. The design of the process flow is also critical to this last point and requires definitive hands-on management for proper collection and development of the ideas. The opportunity also exists to continue expanding and developing the Simultaneous Engineering checklist, which should tend to consolidate into a set of requirements that transcend each phase of the new product development process.

3.2 IDEA GENERATION DESIGN PROCESS

After it has been determined, from the strategic plan and market research, that new product ideas are needed, the next step is to generate them. The process consists of creating an atmosphere where key people inside the company, along with important customers, suppliers, and other stakeholders, get together and come up with new concepts, plans, customer groups, and so on. The format for creating these new ideas has to be well thought out in advance, in order to preserve the open atmosphere and to allow for a free exchange of thoughts, without applying critical judgment. The ideas are then systematically collected for later screening and evaluation.

This is one of the most difficult concepts to implement. Most people involved in high-level meetings, making decisions of great magnitude, have trouble not critiquing new ideas when they first hear them. It is human nature, particularly when the dynamics of a new situation are presented, to want to act on it immediately. More often than not, new ideas are rejected out of hand without a proper analysis. If the idea seems viable, the tendency is to instantly approve it and move the program forward, as soon as possible. Management needs to establish the "no judgment" ground rules up front, and continuously monitor progress so that the "critics" do not take over the session and kill off all of the potential new ideas, before they have been later well thought out.

The venue for this activity is equally important; a site away from the normal job location should be picked, one with limited interruption potential. The organizer should be promoting the idea of committee members divorcing themselves from the everyday problems of their jobs, to concentrate on what the company needs to fulfill its objectives for profitable growth, quality service, and customer satisfaction.

The design of the idea generation process needs to consider several factors implicit in the term "consumer buying behavior," including the cultural, social, personal, demographic, and psychological.[1] These factors should be relied on when developing detailed plans for administering the sessions for generating new product ideas. The research surveyors need to put themselves in the place of the interview participants so that any potentially sensitive issues affecting cultural issues, such as race, religion, gender, and political beliefs, can be screened out or reworded to avoid possible embarrassment or other problems. Short of any negative assessments, this action may also offer insights into the characteristics of these groups, which may help in the formulation of the questions.

Ethical considerations need to be included in the plan, since some focus group participants may not enjoy being observed or videotaped from behind one-way mirrors. The plan should also address downstream concerns such as potential advertising campaign themes, warranty, and service and repair.[2] It is also extremely important for the facilitator to offer alternative choices, in case the focus group gets stuck in a "endless loop" where no reasonable exit exists. For example, suppose a focus group experiences five different product configurations, each designed to perform the same function better than the existing products. If the group cannot collectively decide on a preference for product attributes, the existing products might

[1] Kotler and Armstrong, p. 144.
[2] Ibid. p. 155

provide suitable alternatives, if they are not obsolete. There certainly are times when new product designs are rejected as being too radical or unjustifiably complicated. Simple solutions may still have a place in marketing.

Business buying behavior may pose a completely different problem for researchers. This situation requires a different approach than that developed for consumers. The opportunity to utilize technical data more fully is an obvious advantage. Even here, selective understatement of product performance may prove beneficial when engaging business customers to develop future requirements.[3] For assembling focus groups, identification and inclusion of all members of the buying chain is essential for getting representative and innovative new product ideas. It fulfills the premise that all members of the chain have experience with the current product that could be beneficial for any new product offering.[4]

Brand name research is another area for idea generation. New methods for identifying consumer brand recognition have been the subject of marketing studies for some time, concentrating on the effects of packaging, discount coupons, and private-label competition. Business buyers and re-sellers often treat purchases strictly on an economic basis, with price and delivery as the foremost buying factors. This commodity approach makes product differentiation not only difficult but often impossible. New strategies for increased micromarketing, (i.e., regional differentiation), are recommended as countermeasures in this case.[5]

3.3   NEW IDEA GENERATION TECHNIQUES

Several processes exist for idea generation, the most familiar being brainstorming. Before analyzing this well-known technique, it may be beneficial to examine several others that are more specific. Product feature mapping is utilized to compare the desirable, usable, or valuable features, as perceived by customers, from the market research. Generic product features could include convenience, speed of operation, styling, color, size, power/fuel consumption, quality, cost, among others. During the analysis, each feature is given a numerical range (e.g., 1–10, lowest to highest), and the features are plotted on a two-dimensional grid, with the company product along one scale and competitors along the other. Individual competitors

---

[3] Ibid, p. 166.
[4] Ibid, pp. 180–181.
[5] Ibid, p. 236.

or a composite competitor valuation can be performed, depending on how the variability of the features' ranking between competitors.

After ranking comes benchmarking, to determine the best competitive position for each feature. The features themselves do not necessarily need to be ranked or prioritized, unless there is direction provided by regulation, public policy, company policy, industry standards, or institutional guidelines. One such directional ranking could give product safety priority first followed by environmental impact, energy conservation, mass, size, and so on. Alternative rankings could be based on market wants such as quality, speed of operation, convenience, special ordering capability, and availability.

Market segmentation could also produce a different set of scenarios for ranking, and may need to be studied. Geographic or demographic differences can produce some unusual or unanticipated product applications, based on historical preferences, lack of general product exposure, lifestyle, or need. Sensitivity to these observations in the market research may provide significant advantages for product differentiation opportunities later on.

A natural progression following benchmarking is attribute analysis and extension. If the above comparison of features between company products and competitors show trends where the company product ranks significantly above or below any competitive offerings, the reasons for this phenomenon must be explored, whether or not improvements can be made to change the situation. A simple example would be noting that a competitive attribute perceived as important in all areas, but offered on the company product line in only a few geographic areas, could be extended to all areas with only minor changes in product catalog and marketing literature, manufacturing scheduling, supplier coordination, and shipping arrangements.

There are idea generation techniques that consider even more specific product issues. Treating the product as part of a system is one method. Most product functions can be viewed as steps in a series of sequential operations that need to be performed in order to accomplish some overall objective. Those operational steps both upstream and downstream of the product need to be analyzed for potential improvement opportunities for the product in question. It may become obvious that the entire system is in need of improvement, which could generate even more possibilities for a complete series of new products.

Strengths and weaknesses can be either perceived or real depending on the customer's ability to observe and interpret them accurately. Strengths and weaknesses may not be mutually exclusive or have equivalent effects on the customer. For example, a new vehicle design may be perceived as fresh, original, and dynamic. However, the company's product history may also include multiple service, repair, and warranty problems. The styling may appeal to enough new customers to initially offset the negative impressions from existing customers. This does not mean that efforts should be relaxed in solving the weaknesses. In another case, new styling may not always be perceived as new, but as conservative or "copycat" (i.e., copying another similar product). Product quality is not always recognized by the customer. Certainly, features such as lack of defects, lack of operational glitches, and ease of instruction interpretation, are all strengths that should be pointed out to customers in advertisements.

Product problems may not be the same as weaknesses. A problem with a manufacturing process may be contributing to a high-cost condition, without affecting the customer. On the other hand, an operational problem could only be affecting customers who experience it after owning the product for several months or years, since inside the company, only new versions of the product are targeted for quality auditing or control checks on the assembly line. In this example, if power consumption or battery life is worse than that of the competition, customers might experience it before the engineers inside the company do. Reducing the power requirement or increasing the battery size may not solve the problem if the product design is obsolete. A complete redesign may be required but not feasible in the time allotted for development. This case may need to be completely rethought and a different new product strategy developed.

Gaps or niches in the product lineup may also be evident from the discussion with customers or the analysis of competitive models. This is not to be confused with the line extension theory described above. Gaps are holes in the market where the company simply has nothing to offer to counteract a competitive action, either deliberately or accidentally. A niche is a market segment not covered by any product. Both conditions need analysis to determine whether any effort should be allocated to their development. Frequently, the company has already considered possible action and rejected the proposal as unprofitable or unfeasible. However, that action may have taken place as the result of different circumstances and may need to be reevaluated.

Finally, brainstorming should be considered as an idea generation technique. Ground rules are needed up front in order to provide a focus and a level of containment, so that random or inconsequential thoughts are excluded, without correction. The objective to be stressed, repeatedly, is that collecting the ideas is the primary function and the evaluation will be performed later. Beyond this control point, guidelines for topics can also be presented, although this could be interpreted as a limiting factor that might inhibit the spontaneity of the exercise.

3.4   AJAX MANUFACTURING COMPANY

The third meeting of the new product committee took place about two weeks after the last survey questionnaire was completed. Prior to the meeting, the data was tabulated, analyzed, and distributed to committee members for review. At the session, the ground rules were laid out and explained carefully, so that there would be no misunderstanding about criticism or unintentional screening. The marketing vice president acted as the facilitator and scribe.

He initiated the discussion by showing a slide listing the top 10 problem areas of the model E/D-12. He noted that neither Ajax's selling price nor dealer discount structure were among the major issues. The research manager reported that the three main competitors were priced from 35–37% higher than the E/D-12, but all three had more capabilities, including portability, range of measurement and data capacity. He started a list of general issues on a flip chart and put the above three capabilities at the top of it. Then he went around the room and asked everyone to think about the list of 10 problem areas and to come up with a corresponding list of solutions.

The first item suggested was a handheld version of E/D-12. Others warmed up to this idea by suggesting adding a stationary fixture that would still permit bench measurements. Another idea was a separate readout that could be fixed to the handheld "pilot," for portable readings, or attached to its own stand for bench work. The rest of the ideas were as follows:

Laser-based roughness and profile measurement

Rechargeable battery pack for handheld pilot

Magneto-resistive roughness and profile measurement

Noncontact-based measurement system

Universal ball size stylus

Ultrasonic measurement system

Linear variable differential transformer (LVDT) measurement system

Statistical comparative measurement system

Handheld probe only

Universal measuring instrument

Several other ideas (e.g., to start a completely new business in another field) were not included in this discussion due to their complete separation from the measurement and instrument business. The above suggestions were not analyzed right away, but were allowed to be absorbed by the committee for a few days prior to the screening exercise, which will be explored at the end of the next chapter.

4

SCREENING AND
EVALUATION

4.1 INTRODUCTION: SELECTING THE BEST
NEW PRODUCT IDEAS

The screening and evaluation step is an introductory part of the product
planning process, which is mainly the subject of the next chapter. The reason for concentrating on a particular subset is the result of its relative importance to the balance of the new product development process. Another
is the need to perform an in-depth analysis of certain rating factors developed over the years by experts in the new product development field.
These factors include the need to rate three major qualitative as well as
quantitative factors:

  Marketability

  Manufacturability

  Profitability

A checklist of extensive items is compiled as a result of the thoughts
collected during idea generation, as described in the last chapter. Then,
during the analysis, each item is considered by itself and in conjunction
with others, to develop a rating based on the above three factors. Judgment
is required to determine the combined effects of some issues, which can
be extremely complex. In other cases, the lack of one of the three elements
could negate the entire proposal. However, it is perhaps fortunate to be
able to rule out an obvious or highly unprofitable venture in the beginning,
rather than arriving at the same conclusion after a few weeks' expenditure
of developmental costs and lost time.

This last point, then, is in the true spirit of the new product development process. It is also the perfect time to bring together all of the Simultaneous Engineering principles and focus carefully on each issue for potential problems and opportunities. Unfortunately, there is also more of a tendency to not pursue a new venture due to the inherent risks, in spite of potential rewards, than take a chance on a questionable or doubtful new product idea. The following plan carefully describes how to identify those risks that are worth pursuing.

4.2   RATING CRITERIA DEVELOPMENT

The new product committee organized during the strategic planning stage now convenes to discuss the new ideas that were collected previously. Their objective is to consider whether any of the new ideas have merit, from three key perspectives: marketing, manufacturing, and financial. These can be thought of as equivalent to the three legs of a stool, unstable and unworkable if any one of them is deficient or missing, despite the strength of the remaining two. This criterion has been developed over many years of observation and through the analysis of records, which is indicative of successful products through their history of long life cycles, profitability, and sales. The characteristics of these successful products can be traced to either creating or filling a preexisting need, one easily manufactured and with a sufficiently long life to recoup its initial capital investment and turn a profit comparable to or greater than other products in the same market.

Since product cost is prominently featured in each of the three areas, it is important to understand how and why this has occurred. Virtually all factors involved with new product development can be related to one type of cost or another. Direct manufacturing, product development, and capital investment costs are probably the most noticeable and easiest to calculate and analyze. However, the costs implicit in doing one thing over another, and with not doing something or not anticipating some future competitive action or reaction cost, are much more difficult to predict. Thus, the concept of Simultaneous Engineering was developed to respond to these more elusive issues. Hence, also, is the need to develop a rating system that can address all of the concerns, both obvious and hidden. The rating criteria are explained below, and there are examples also in the mathematical model development chapters.

Initially, each of the three sets of criteria (marketing, manufacturing, and financial) can have equal weight, unless the company's particular situation dictates an alternative stance. Since, in a generic situation, the company already has an established product line, the starting point can be comparing the new ideas to the existing ones. The point needs to be stressed that, because the new product idea may have a potential impact on the existing product, the comparison should take into account how the new product could be designed and manufactured, relative to the existing one. Some of the generic product screening criteria are listed below.

4.3   PRODUCT-BASED MANUFACTURABILITY CRITERIA

The following list of questions is used to determine if a new product can be manufactured per the ideas outlined as a result of the generation process in Chapter 3. Note that the ideas may not have been completely thought through:

Fits within the company's current capabilities of:

  Existing product lines
  Physical plant space
  Available equipment
  Tooling and fixtures
  Personnel competence levels
  Engineering talent
  Technology
  Management skills

Fits within suppliers' current capabilities of:

  Existing product lines
  Physical plant space
  Available equipment
  Tooling and fixtures
  Personnel competence levels
  Engineering talent
  Technology
  Management skills

Are areas of uncertainty in the new product that could impact or affect its manufacturability, such as:

Component fabrication

Assembly sequence

Calibration of operations

Tooling capabilities

Advanced or unproven technology

4.3.1   Simultaneous Engineering Criteria

The following questions relate specifically to design and engineering parameters that the new product will have to meet or exceed. The concept may not yet be sufficiently advanced to answer all of the questions but asking them at an early stage may uncover an issue, problem or opportunity that should be addressed before any further work takes place.

The manufacturing complexity is better or worse than the competition.

Disassembly sequence for recyclability is better or worse than the competition.

4.4   MARKET-BASED CRITERIA

The market impact assessment of a new product considers what needs to be accomplished in order for the product to be sold. Identification with specific and general buyers, existing product distribution channels, promotional plans, advertising budgets, relations with dealers and sales agencies, inventory and pricing policies, and service and warranty can all have positive or negative effects on the new product, as well as on the existing ones.

Potential competitive actions also need to be analyzed. What if the new product cuts into an established competitor's "turf," causing an unanticipated retaliation? Some of the more important marketing criteria are listed below:

A real need or want for the new product has been established.

The new product represents a breakthrough concept.

The new product is too new for acceptance.

Describe or highlight the new product's main and unique features succinctly.

Describe its competitive advantages.

Potential competitive reaction to the new product.

Easily copied by a competitor; affordable action to counter this move.

Attributes that will discriminate the new product versus the competition.

New product as an imitation of an existing product.

Market share splintered among many competitors.

If market considers the new product as a commodity, can it compete on price alone?

Product creating its own niche.

Appeal of the new product to any other target markets; risks of including these other segments.

Potential for new market segments if the new product keeps evolving.

Alternative products that are functionally equivalent to the new product.

Distributor profit selling the new product.

New product cost or price range.

Projected market share price elasticity.

New product's expected market share.

Expected growth rate.

Ultimate market position.

Potential for cannibalization from other company products.

Effectiveness of the promotional material.

New distribution channel required.

4.5   FINANCIALLY BASED CRITERIA

The financial analysis requirements for a new product is based on the philosophy that it should be able to pay back the original development, marketing, and other direct costs, early enough in the life cycle that the remaining profit cash flow can be reinvested in new technology, facilities, and research for the new product's eventual replacement. Initially, the financial issues center around the profitability of the new product and how the profit will be utilized to pay back the investment incurred in its development. Then, later on in the life cycle, profits can be utilized to offset fixed expenses or variable costs from other products. Finally, provision needs to be made for the new product's replacement and the capitalization required, if there is sufficient funds left over from other uses. Currently, product life cycles have become sufficiently foreshortened that new product funding may have to come from the sale of additional equities or other financial instruments.

4.5.1   Calculations

A list of the key financial issues in new product development follows:

Direct Costs

Purchased components
Raw materials
Fabrication and assembly labor
Advertising and promotion
Sales
Distribution
Warranty
Service and repair

Indirect Costs (Overhead)

Facilities (leased, rented, or financed)
Utilities (heat, light, power, water, telephone)
Operations (administration) and management
Effect of future cost reductions on the marketing position of the new product
Price of the new product based on a combination of both its costs and the market.
New product's gross profit and contribution
Potential breakeven point
Projected total annual profit
Expected return on investment; on sales; on assets
Cash-flow considerations
Expected life cycle
Product's capital requirements

Other issues may also become significant if the product development process continues, but the above list is an approximate screen for initial consideration. There may be additional requirements for specific industries, depending on how the industry handles financial transactions and fund utilization. Most firms probably prefer to finance new product development from internal reserves or retained earnings from earlier products. However, large projects or new technology may require external capitalization in either equities or bonds. Quite often, stock issues are tied to a company's overall financial needs, not just to innovation. Bonds tend to be utilized for large longer-term, quasi-public projects such as power plants, dams, and railroad expansion. Mergers and acquisitions can generate cash for new product development if existing assets are liquid and the market timing is right.

As was demonstrated in previous chapters, the above effort can be translated into a Simultaneous Engineering process. In this case, it consists of the above checklists to determine product manufacturing feasibility, marketability, and financial viability. Criteria from all three areas must be addressed and reconciled in order to have a successful new product venture. It will probably not be possible to positively answer all of the above criteria, but the more that are answered positively the greater the insurance that the new product development process is on the right track.

The next chapter explains how the criteria from the checklists gets translated into real product specifications, how the various development teams are organized, and how timing and other measurement devices are utilized to insure success.

4.6   AJAX MANUFACTURING COMPANY

The new product committee had five working days to individually review the new product ideas generated at the last meeting. Many members had reservations about one or more of the brainstormed concepts, but said nothing in the meeting or afterwards, in keeping with the guidelines accepted by the whole committee. During the following week, several committee members performed their own analysis on the product development costs, potential wholesale and retail pricing policies, design and engineering problems, service, warranty, and so on. Committee members were asked to confer with other committee members, not on any conclusions, only on technical points. Despite this restriction, most of them had formed, on their own, very definite opinions as to the feasibility of several new technologies suggested in the brainstorming session.

The marketing vice president opened the meeting by requesting comments on the first item on the list, a portable E/D-12. The manufacturing director asked why the firm should waste its resources on an obsolete technology. The research vice president agreed, and further stated that even the most popular competitive instruments were also obsolete. The president asked why customers seemed to be satisfied with these old units. Two of the five major transportation industry manufacturers, who were on the committee, discussed the question briefly between them. One of them said that most of the quality assurance measurements were being performed by people not yet trained on the newest equipment. Early tryouts had not been successful because the time prescribed to perform the measurements was inadequate, or it took too much time due to lack of experience. This time lag was thought to be a problem for at least one more year, due to a continuation of new instruments being evaluated. No decision had yet been made about preferences, as it seemed the inspectors and operators each had their own opinions. The companies were not pressuring the union to achieve a consensus, since the issue was considered part of a training course for all employees, for continuous product improvement. Every month, new measuring instruments were being introduced into the plants, with training seminars given free by the vendors every week, both during and after regular working hours. However, there did not seem to be a coordinated program to explore measurement concepts from a strictly educational standpoint, since all the courses were designed to utilize the unique features of the instrument being pitched.

Some of the more exotic technologies under investigation by the teams were, admittedly, geared toward laboratory work, where ultraprecision was not only desired but necessary. The laser and noncontact or ultrasonic systems fit this mode of operation and therefore were not practical for the shop floor. Also, some of the more sensitive systems could not operate in the shop environment where excessive noise, dirt, heat, humidity, and vibrations from heavy machinery, such as stamping presses and forges, caused calibrations to drift out of specification limits on a regular basis. These last points were brought out by the supplier of castings to Ajax.

One of the other customers raised the point that Ajax instruments were not overly defective; in fact, their reputation was generally good. The Ajax quality control manager had investigated the problem with binding and loose pilot rods. It resulted from a particular shipment of semifinished material that warped due to improper heat treating, but was not detected during final machining, polishing, and assembly. Final inspection did not catch the problem because the warping seemed to progress after shipping.

The shipping schedule was particularly tight and the units in question were completed and shipped quickly, without the usual one- to two-weeks' run-in after assembly because of unusual customer demand. Incoming inspection at the plants had uncovered the problem, prior to the release of the units to the production workers. Replaced pilot rods were rushed through production and hand-carried to the plants, installed, and inspected by Ajax factory service technicians.

The other reliability problems mentioned in the idea generation meeting came from random failures of motors, switches, relays, and readout displays. The failure rates were within $\pm 10\%$ of the average of all similar instruments tested by Ajax lab technicians. The suggestion may have been directed at improving the entire industry and not just the E/D-12.

The central issue was next debated; i.e., what was the best overall technology for the measurements required? The resulting debate ranged from a simple and quick method to complex and involved one, with no clear-cut advantage of one over the other.

Then it was pointed out by the engineering director that one of the more simple technologies, magneto-resistive, could handle both roughness and profile measurements without changing anything except the scale on the readout device. This was due to a new idea proposed for a pivot lever arm, along with electronic compensating, which would allow both roughness and profile measurements, to be performed by the same probe, without stylus replacement or recalibration. This meant that a machinist could check the roundness and roughness of a workpiece with one measurement.

Because of the experimental nature of the technology, it was agreed that a dual effort would be pursued. The linear differential variable transformer (LVDT) concept would also be developed, along with the magneto-resistive method, to the point of mass production, to verify which had the most potential for long-range growth. This was in spite of the problems of drift, associated with the LVDT technology, that typically required more complex electronic control circuitry.

The consensus of the new product committee was to divide the group into two teams, each pursuing a different goal, to develop these concepts into production-ready programs. At the same time, the universal measuring instrument concept was temporarily shelved until the new combination profile/roughness instrument product concept could be proven.

The next chapter will show how Ajax went about developing the dual programs.

5
PRODUCT PLANNING

5.1 INTRODUCTION: TRANSFORMING CUSTOMER REQUIREMENTS INTO NEW PRODUCT SPECIFICATIONS

In this chapter it will be shown how a company manages the conversion of a concept into an actual new product. The effort requires the resources and coordination of all major stakeholders, applying as much synergy as possible to solve common as well as unique team problems. Achieving a consensus on the product definition may be the most difficult problem to solve, since many solutions are usually available.

The product planning process consists of several steps, an initial product concept, concept development and testing, marketing strategy development, business analysis, initial product development, test marketing, and commercialization. The steps are not completely sequential, but overlapping, as has been shown in previous chapters. An initial product concept can be conceived at any time, even before the development of a new product corporate business strategy. As previously shown in the market research arena, customer needs and wants can provide the stimulus for a large number of new ideas. Some are selected for immediate pursuit while others are relegated to a secondary status. This process follows a plan that concentrates efforts to develop the concept(s) voted both most likely to succeed and offering the most potential for short- and long-range profitable growth.

However, it should be remembered that new ideas do not always work out, that those thought to be the most promising may, indeed, be the most faulty, after all factors are considered. This is why many companies should have several new product teams working either in friendly competition or on projects relegated to a less stringent development schedule due to a

56

lesser priority. These secondary projects could become primary programs very quickly if technological breakthroughs or failures are achieved in any one of the ongoing programs.

5.2   DEVELOPMENT OF A MARKETING PLAN

Following through on the above premise, once a product concept (i.e., a detailed version of a new product idea), is translated into terms that relate to the consumer, it needs to be developed and tested.

Concept development could resolve itself into either one or many ideas, each requiring further development. This situation is where Simultaneous Engineering would offer significant benefits over a consecutively executed program, providing shortcuts in each developmental phase. When the project completion handoffs between groups are eliminated, the project necessarily becomes shorter. This naturally assumes there is a good working relationship between the various groups (marketing, design, engineering, and so on).

Concept testing involves another specialized branch of market research, one in which customer wants and needs have been thought through and formulated into a solution. This formulation remains to be tested to make sure the concept is being carried out correctly, which should be continually checked throughout the later phases of the entire new product development process.

Testing consists of an initial development step that adds sufficient content to the idea, which can then be explained or presented to selected customers for feedback. The test questions that need to be answered include how the new idea compares with its closest competitive unit or existing product from the company. Any and all issues should be explored for answers, including these:

Styling
Color
Size
Pricing
Convenience
Operational safety
Functional performance
Environmental impact

The methods utilized in addressing each issue may range from the way in which the survey questions are asked to requiring an explanation from the respondent. However, perceptions of a new product concept can be radically different, depending on the recipient of the information. Even when perceptions are similar, their relative importance can vary by issue as well as by the recipient. Attribute ranking is one method for establishing the significance of specific features. It should be an integral part of every new product concept survey.

Based on the concept test results, a marketing plan or strategy is next developed. This is the essence of the product planning process element. The target market, product position, sales volume, profitability, and market share should be derived from the initial product concept study. The forecasts may not prevail, but the plan needs be in place to provide the necessary details. Longer-term factors also need to be defined in as much detail as possible, including advertising budget, pricing policies, profitability goals, and the marketing mix with other company products, among others.

Next, a business case analysis should be performed utilizing many of the questions that appeared in the financial checklist in the previous chapter. This exercise is to assure senior management that the new product will meet corporate goals for profitability and growth. If there is a problem at this juncture, the project could be canceled unless a new strategy can be developed. This is the final checkpoint prior to initial product development and/or limited production, which is a logical decision time. After this, any major changes in the program are going to be expensive, since manufacturing tooling is involved. A sample product planning proposal form is given in the Appendixes, and several filled-in versions are given in Chapter 11 for the fictional National Motors Corporation.

5.3   TEST MARKETING

Test marketing may or may not be required, depending on senior management's confidence level in the new product. The advantages of early customer contact and feedback must be compared against the obvious risks of a loss of secrecy, competitive reactions, and potential sabotage. Several types of consumer product test markets are available, ranging from the standard open market to store-controlled environments to simulations, in which customers can sample new products or competitors in a store-type setting. Purchase data from the bar code scanners are available in all three cases. Market subsegmentation may also be attempted in a bid to further learn about targeted buyers.

For business-type products, other forms of test markets are available. These range from trade shows and industrial expositions to private display rooms to private utilization situations, where selected customers try out the new products in their own facilities and report back any problems or other observations to the manufacturer. Security, competitive reaction, and retaliation can also result from this move.

Initial product development is similar to early prototyping, except final designs have not yet been completed. Functionality is the main purpose of this phase. In the electronics industry, this used to be referred to as "breadboarding." This step resulted in temporary circuits constructed on small Peg-Boards, with components crudely mounted and soldered together, without regard to final positioning or other packaging considerations. Other products may utilize existing shells or external coverings from obsolete products and loaded with all new components.

The automobile industry calls these early prototypes "mules" if they borrow extensively from earlier production units. Expediency in development is important, as is disguising field test vehicles from spy photographers. Early test results from these "expendable" units can provide valuable information to assist in completing final designs.

5.4   COMMERCIALIZATION

After the initial concept has been proven, the final decision is commercialization, that is, whether or not to proceed with full production. This milestone should not be approached lightly, for the result could prove either decidedly good or disastrous to the firm. All of the past Simultaneous Engineering efforts culminate in the actions that lead to this juncture; whether to go forward or backward. This is because, if the company chooses not to proceed and the competition does, the company could be in a recession. It may move forward and fail, backward and fail, or stand still and fail—or it may succeed. It depends on the rapid and correct blending of all of the above talents. It also depends on what the competition thinks will happen and what they do, in response to the company's actions. In any case, after the decision is made, the entire situation may change; the company needs to be flexible in its response to it.

It is also important to note that companies are utilizing more computer-aided design (CAD) for initial product concept development and simulation, which not only has reduced the need for early prototypes, but also allows some products to be put directly into production without a great deal of field testing on the finished version.

While coordination of this particular process step is usually handled by senior management, in some cases it is preferred that the new product committee be in charge, with subcommittees comprised of membership from such disciplines as marketing, timing, design, engineering, purchasing, manufacturing, and service.

Since this work requires multidisciplinary skills, selection of committee members should have one of the highest priorities. Unfortunately, in some firms, assignment to the new product committee or one or more of the above subcommittees would not be considered a priority and would not receive the support it deserves. The kind of crossover skills required range from current product engineering to assembly, vendor relations, sales experience, service familiarity, and so on. Due to current trends in worker specialization and downsized companies, fewer and fewer employees have these crossover skills. In the future, successful companies will have to provide opportunities for this kind of training inside their own operations.

An alternative would be to rely on suppliers, retirees, subcontractors, and consultants. Project timing specialists are even rarer. Fortunately, there are extensive software programs for plotting Gantt, PERT charts, and project management tools to provide another alternative for shorthanded firms.

The key factor in this context is project ownership. While committees are good ways to distribute workload, assign responsibility, and assess accomplishment, overall project coordination usually requires a responsible leader with sufficient authority to handle disputes between groups, make decisions on the basis of group or majority consensus, and maintain development schedules, particularly when other product priorities are demanding a higher priority or budget allocation.

New product innovation frequently requires risk taking, often when a new or unproven technology is involved. This action has to be supported and, even if unsuccessful, rewarded by senior management. Committee members should also assume their share of project responsibility and reward. However, most companies are sufficiently shorthanded that no one has just one assignment or job duty. Most employees are constantly juggling multiple projects, bosses, work flow, meeting schedules, and deadlines.

5.5   GENERIC CASE EXAMPLE

In the generic case, assume that a company has formulated the idea for a universal measurement product. The strategic business plan calls for the development of a "breakthrough" product that will establish the company as the technological leader in this field. A strategic plan has been developed and approved, authorizing senior management to investigate whether the measuring instrument market is ripe for entry and whether the "universal" instrument is the right product for that market. The results of surveying customers having experience with current products have been analyzed. The various subcommittees have had their initial meetings and have reported. The new product committee has conducted idea generation and screening/evaluation exercises. Now it is time to develop specifications for the new instrument that will satisfy all requirements to date.

The design subcommittee has submitted preliminary sketches of the "universal" instrument. The engineering subcommittee has submitted estimates for costs and timing along with operational interface and system requirements to do a feasibility study prior to the project, for several major market applications. The manufacturing subcommittee's report was not as positive as the previous two. Feasibility is not assured, based on the technology required to fabricate the new integrated circuit (IC) design. The scale of miniaturization was feasible for the laboratory environment, but unprecedented for the industrial market, except for aerospace applications. The technology is not considered workable from inside the company. No competitor has ever attempted manufacturing such a small device before, although several patents have been applied for but not issued. A supplier needs to be found with the necessary expertise and willingness to gamble on this venture.

The purchasing subcommittee had the task of finding the appropriate vendor who may already have an aerospace contract for this component. After consulting a list of approved aerospace suppliers, three were selected and contacted concerning the subminiature IC. Only one vendor is supplying a major military contractor with a small number of devices similar to what is specified for the company's "universal" measuring machine. A "minor redesign" would make the IC fit the universal application, but the changes in manufacturing and tooling would delay the project six months. The unknown, at this point, is whether the circuit would work in this particular application.

Based on the latest market research, the universal measuring device is designed to be applicable to several miniaturized instruments, some that could be handheld. Some of the applications include surface contour and finish, temperature, pressure, voltage, current, and resistance. Separate input and output modules are plugged into the device with small quick-release fasteners. The IC device and input sensor/output display could be attached to the wrist with elastic or mating friction straps, or carried in a holster with a flip-up display, to permit free hand movement.

A different type of sensor/display unit would be designed for each input/output function desired, along with corresponding sensitivity, multiple-range, critical damping, and other compensation circuits. The main advantage to this technology would be increased portability and nearly universal applicability. The integrated circuit could then be designed into many unique instrument applications, over many models, years, and so forth. If it works, it will render obsolete virtually all measuring instruments overnight. Later, the potential supplier tells the purchasing subcommittee that the military application is a high-altitude, high g load sensor and computer that provides input voltage to the rate gyro display and acts as a damper to limit input to the attitude indicator at high rates of acceleration and deceleration, to prevent voltage surge damage to the instrument.

The new product committee now has a dilemma. Although several subcommittees have submitted positive responses, the manufacturing subcommittee will not support this effort and has further challenged any outside company to make it work. The fit of the military application seems to require an extremely long "leap of faith" to become the universal instrument. Senior management wants to review the specifications as soon as possible, in order to discuss the universal instrument with some key company and institutional investors. The product safety subcommittee has just issued a report in which doubt was expressed as to how the owner's manual is going to define safe operating conditions, when so many different parameters can be measured by the same instrument, with only a few slight input modifications. A primary concern was expressed by the product safety subcommittee when measuring high temperatures or pressures. Also, a paper failure analysis shows that a large backup shunt resistor is necessary to keep a failed IC from contributing to a life-threatening short circuit, when high voltages are being measured. These new issues have added to the problem and further delayed the decision, but not for long.

The decision is made to cancel all further work on this proposal. The reasons given to senior management include concerns over operator safety and manufacturability. Obviously, senior management has heard about it and reacted prematurely to "get things rolling" before the competition steals the idea right out from under them. The product planning decision function has concluded that there are simply too many unknowns and potential problems for this project to overcome with profitable costs and proper market introduction timing.

5.6   AJAX MANUFACTURING COMPANY

Ajax production and research department technicians were assigned to each development team. Together with the engineers, the teams managed to build several rough, early prototypes for evaluation by selected customers, suppliers, and senior management. This process took two months of intense concentration from virtually all Ajax employees and assistance from suppliers.

Team 1 was working on the magneto-resistive concept. This involved manually moving the probe, with a stylus on the outer end, over the work surface, at any particular speed, which, in turn, moved the other end of the probe rod through a set of pivot points. A small magnet, mounted to the inner end of the probe rod, passed over another set of magnetic arrays, inducing a voltage proportional to the arc of movement and to the roughness or profile being measured. The output voltage remained on the display until it was zeroed out. Because most machined surfaces are sinusoidal, the device measured the total change in height, from peak to valley, depending on the diameter of the stylus. Starting with the smallest stylus, with a diameter of 0.0001 mm, would assure the operator of gathering the most information. However, in most cases, the operator would have some range of roughness or profile in mind, based on past experience. Larger styli could then be substituted if the roughness level warranted it. Another advantage of this system was that virtually no contact pressure would be required, since measuring speed was not a factor. (Earlier designs required a designated measuring speed to control stylus skip.) The one-pass system should also minimize any surface degradation due to the stylus.

Team 2 was progressing nicely with the LVDT concept, although setbacks did happen during the machining of the transformer core. Fit tolerances were much tighter than with the E/D-12 probes. After a short learning interval, factory technicians produced the pieces at the required specification. This system employed a moving coil that generated a magnetic field and voltage, in the outer or secondary windings, proportional to the linear movement of the coil piece (attached to the probe pivot bar). The range of coil movement was only about 10% of that from the magneto-resistive device. Thus, different pivot bar arrangements were necessary to align the vertical motion of the stylus along the work surface to the maximum coil movement expected. Again, some advance expectation of the surface roughness might save time in the setup but would not be not mandatory. A stylus changeover was neither complicated nor time-consuming. However, flexibility of the device allowed it to be converted to a completely different instrument, for profile measurement. With the pivot bar adjusted to an asymmetric position, surface geometry could be measured. This meant that a second probe (i.e., larger diameter stylus) would be unnecessary, except to prevent damaging a soft surface, such as aluminum.

The same readout device could be utilized for both the LVDT sensors and the magneto-resistive sensor package. Packaging and power requirements were not equivalent, since the LVDT utilized a small can motor to propel the probe rod. Batteries and/or a power supply was necessary for the unit to operate. The magneto-resistive unit did not require either a motor or a battery/power supply, but it did require a cable connecting it to the readout display. The LVDT could accommodate a small readout along with the connection for the display.

At the same time as the early prototypes were in development, the marketing subcommittee was creating a product plan, complete with a strategy for targeting midlevel customers, such as tiers 1 and 2 automotive and aerospace suppliers. Since both technologies were still in the running, two separate plans had to be compiled, although some elements were common to both. A travel schedule was prepared for the following two months, in which representatives from both teams would make separate presentations and demonstrations of their respective instrument concepts. Then, customers could try them out on the factory floor and compare the results. In-house tests by Ajax technicians showed that the data from each was equivalent. Furthermore, measurement variance calculations showed that the tolerance band was within the specification adhered to by the customers.

Minor operational problems were a constant source of irritation to both teams, but due to different reasons. Team 1 had reliability problems with the pilot pivot-arm mechanism, which kept exceeding its adjustment tolerance due to nonlinear spring tension in the design. Team 2 had stability problems with its electronic circuits, due to drift at startup and operating temperature. Compensation for temperature changes was a stopgap design of three manual adjustments, instead of an automatic system, which would be standard for the production design. Finally, the tests were stopped until the redesigned components and circuits could be installed in their respective units and checked for effectiveness. Then the tests continued.

Test results confirmed that, at first, customers seemed reluctant to rely on only one pass for data collection. Since the same value was recorded in succeeding passes, they thought the machines were not working properly. The lack of a motor kept the weight of the magneto-resistive unit very low. One user commented that it seemed flimsy; that there was nothing inside the shell, compared to other miniature motorized devices. However, when switching over to measure the roundness of a cylindrical shape, the operator simply mounted the device to a fixture and rotated the workpiece around once to obtain the reading. The microcomputer in the readout automatically scaled the output correctly. Most users were impressed by the speed and accuracy of the readings.

Even team 2's device, the LVDT, which also did not rely on tracking speed for accuracy, received responses suggesting the operator was not aware of its performance. On small surfaces, one had to observe closely to verify that the pilot rod did move the stylus out and back, for the stroke could be as short as 10 mm. The tolerance of the backlash adjustment for the feed screw attached to the pilot rod was somewhat problematic. On numerous occasions, the pilot rod appeared to have skipped or interrupted its traverse during the direction reversal while the clearance in the threaded end of the pilot rod was taken up. A new thread arrangement was necessary, and it was under redesign by the end of the tests.

Both teams received an abundance of data to utilize in the final designs of their devices. The final package for team 1 would need to conform to a more user-friendly hand fit, to permit utilization on virtually all surfaces. Adding mass actually helped the stability of the instrument for holding it on nearly horizontal, curved surfaces.

Team 2 had already identified its major tasks and was well into the activities as the test data was being analyzed. However, the analysis revealed that a voltage spike had occurred over the course of several measurements, all at the same facility. Attempts to reproduce the spikes came to naught, until a study of the electromagnetic protection system showed a problem. A loose ground connection had permitted high-frequency electromagnetic interference to penetrate the display box and contaminate the circuit, causing the spikes.

Based on the test results, both teams prepared their final design specifications and marketing strategies for the final decision of the full committee. Everything was discussed in detail, including the data analysis, customer comments, supplier reactions, technicians' findings, and so forth. In the end, the committee voted 7 to 5 to approve the LVDT program for final engineering design and production, although they strongly recommended continuing with the magneto-resistive project. The main reasons for the

LVDT proposal's acceptance were customer familiarity with similar competitive instruments, and an existing track record of service and repair history. Customers were enthusiastic about the other program and urged Ajax executives to refine the instrument for further hands-on testing and development.

6

PRODUCT DESIGN AND ENGINEERING

6.1 INTRODUCTION: ASSURING THAT ALL DESIGN AND ENGINEERING OBJECTIVES CAN BE ACHIEVED

This is the main focus of the book: where the adoption of Simultaneous Engineering principles should bring all of the previous inputs together so that the product can be designed to be right the first time. Product design is a complex, dynamic combination of art and science. The principal goal of the designer is to create a unique, awareness-enhancing shape or form of a product that is directly related to or complements its functionality. The principal goal of the engineer is to translate the designer's creativity into a body of knowledge that provides the desired functions, in a manner that is both practical and cost-effective to manufacture and to service or repair, and that minimizes environmental impact.

The chapter covers two topics: product design and product engineering. Design activities are those that enhance, define, describe, and develop the specifications and expectations of the desired product into the form and function outlined in the product planning activity described in the last chapter. Engineering activities include those that transform the designed material and specifications needed for both the preproduction or prototype version, and the final or production version of the actual product.

Note: It is important to understand that the expectations of each succeeding phase element in new product development will get interpreted or misinterpreted as an outcome of the organization's efficiency and effectiveness. Mistakes and misinterpretations are as much a part of human nature as any other activity. Minimizing the impact of mistakes and misinterpretations is a primary objective of Simultaneous Engineering. The proposed system seeks to clarify goals and compare results at each phase of new product development, thereby minimizing misunderstandings between all project handoff partners. However, as was also pointed out earlier, there may be changes in the program along the way that require reconsideration of earlier decisions. Some of the changes could be the result of external inputs, market effects, or other stimuli. New design parameters might become mandatory due to legal promulgations, or should be anticipated if new regulations are being discussed, perhaps even for possible harmonization between political bodies. The process should always be left sufficiently open to allow for unanticipated inputs that require sudden and decisive actions, should they be required on short notice.

6.2   DEFINITION OF PRODUCT DESIGN

The design department receives input from the new product committee in the form of a list of descriptions and specifications of the product's functional and nonfunctional (e.g., appearance, styling, and color) characteristics. Other desirable outcomes might include safety, cost, packaging, power requirements, system compatibility, operational parameters, and recyclability. While it may seem that some items are engineering requirements, these specification and optimization processes should be performed in an iterative continuum, with both designers and engineers participating in the analyses and discussions. It is important to distinguish between the two disciplines only as they relate to opportunities for process improvement. Some products are invisible to the customer or hidden inside a larger product. Aesthetics may not be very significant to the buyer, unless the configuration can influence the product's functionality or marketability.

The designer may also consider competitive product features and attributes. Assuming the research has been thorough enough to rank these attributes, the designer can make a preliminary decision regarding overall product size, shape, layout, and function. The product at this point becomes nothing more than a transfer function; that is, it receives input in some form, processes it, and then transfers the output, in the same form or a different form than the input. This "black box" approach can be successful when involved in a project with tight timing and a high reliance on qualified suppliers. Final specifications for the device may not be completely defined until well into the prototype phase. Following this methodology, whole groups of black boxes can be strung together, in series, parallel, or in combination, to form a product "system" block diagram. Such systems have long been useful in aerospace, military, and computer applications for analyzing and developing product functions and responses. The automotive and heavy vehicle markets are also beginning to utilize them, effectively.

6.2.1 Development of Design Transfer Functions

Development of transfer functions for the five industry groups discussed later in the text should provide some insight as to how the designer should analyze the situation in which there is little or no facing competition for a new product concept, that is, no benchmark for comparison.

The five industry transfer function statements could be described as follows:

Automotive: To provide light-duty (8500 lb maximum gross vehicle weight) surface vehicles capable of transporting passengers, cargo, or a combination thereof; in safety and comfort; with minimum environmental impact; in a medium that enhances profitable growth for the manufacturer and all others in the supply chain, as well as goodwill, and other intrinsic, socially acceptable benefits.

Aerospace: To provide air/space vehicles capable of transporting passengers, cargo, or a combination thereof; in safety and comfort; with minimum environmental impact; in a medium that enhances profitable growth for the carrier and manufacturer and all others in the supply chain, as well as goodwill, and other intrinsic, socially acceptable benefits.

Heavy/Off-Road/Military: To provide heavy-duty ground vehicles capable of transporting passengers, cargo equipment, or a combination thereof; on both highway and off-road terrain; in safety and comfort; with minimum environmental impact; in a medium that enhances profitable growth for the manufacturer and others in the supply chain, as well as goodwill, and other intrinsic, socially acceptable benefits.

Electronics: To provide a medium that transports and processes electrical information into a variety of outputs, with minimum environmental impact, which enhances profitable growth for the manufacturer, all others in the supply chain, and end users, as well as goodwill and other intrinsic, socially acceptable benefits.

Consumer Packaged Goods: To provide a medium to transport retail goods to consumers; in a safe, convenient, and environmentally sustainable form that does not detract from that product being transported; that provides profitable growth for manufacturer, processor, all others in the supply chain, and the end-user, as well as goodwill and other socially acceptable benefits.

Without restrictions, the above descriptions, or others, could be utilized to define new properties and opportunities within these industries as well as others. However, restrictions are frequently in place for all of the manufacturers in the above industries. Existing product histories, either within or outside of the category being considered for a new product, carry much influence. Also, constraints on manufacturing capabilities, infrastructure issues, supply-chain limitations, sales organizations, training requirements, among others, are equally important in the decision-making process for pursuing a new product. Ideally, all of the above should have occurred during earlier process phase elements of the new product development, such as idea generation and screening or evaluation. It will be assumed that the list of specifications and desired outcomes, arrived at during product planning, already has been critiqued against available skills, abilities, and resources and the design/engineering process now ready to begin.

6.3 ENGINEERING

In the not-too-distant past, product design and product engineering were distinct and separate activities. Communication and technical problems occurred with regularity, even when management was combined. With an increasingly competitive market, the need to bring out new products with high-quality features, zero defects, on time and within budget necessitated a new product development system.

Simultaneous Engineering can play a major role in design/engineering, through a compilation and analysis of facts learned from earlier studies, to help speed up the new product development process.[1] Beginning with the expected function of the product, designers and engineers must visualize all of the different scenarios that the product could be utilized. This may seem routine, if the new product is basically carried over from the existing one. In order to insure that no new additional uses or applications are being explored, the designer/engineer checklist, shown in the Appendix, should always be administered, regardless of the new product application or previous usage history.

Design/engineering information flow and control will have a major impact on the timing and effectiveness of Simultaneous Engineering activi-

---

[1] Kotler and Armstrong, pp. 325, 326.

ties. Design work and engineering outputs occur throughout the life of the program. Frequency and accuracy of updates are as important as the material being updated. With computer-aided design and engineering programs, the need for periodic reviews, with all affected parties present, may be past. However, the need for new information has not diminished, only changed in form. Communication technology has advanced to the point that remote location reviews, electronic data transfer and virtual reality opportunities abound and are being implemented.

6.4   AUTOMOTIVE INDUSTRY

The major factors affecting new vehicle design include the following, in no particular order:

Compliance with safety standards
Compliance with environmental regulations
Manufacturability (manufacturing complexity)
Warranty and serviceability
Quality and customer satisfaction index
Durability and reliability
Performance or functionality
Damageability and repairability and insurability
Recyclability
Future and unanticipated requirements

All of the above should be considered cost parameters that need to be optimized or minimized simultaneously, if possible. The effort should be combined in team efforts including engineering, marketing, purchasing, service, suppliers, among others, in an iterative process, so that handoff problems are eliminated. Each of the above parameters have several segments that need to be considered, first separately and then combined. Also, there are several cost categories within each parameter that could be analyzed separately, if their relative magnitude warrants it. The cost categories include development, opportunity, component, labor, overhead, and so on, which should all be considered important until a subsequent analysis contradicts this hypothesis.

These cost parameters were developed so that designers and engineers could visualize the relationships between them and their singular and combined effects on the overall vehicle. These parametric objectives have existed for a long time, whether inherent or not. In the past, some of these costs were either buried in a overhead category or not accounted for because the need had not yet been established or prioritized.

Some of the parameters are based on government regulations, such as the Federal Motor Vehicle Safety Standards, the Clean Air Act and Amendments (emission control regulations), and CAFE (the corporate average fuel economy laws—one for passenger cars and a different one for light-duty trucks). All of the above are primarily affected by vehicle size and mass. Others are internal automobile company target values or benchmarks, utilized for marketing purposes including manufacturing, performance and quality improvement, warranty, insurance, and repair cost reduction. There are also those regulators that anticipate a future direction, such as the application of all safety standards to light-duty trucks, harmonization of all worldwide regulations, and near total vehicle recyclability. There are also discussions taking place to harmonize safety standards between Europe and North America, the positive outcome of which is deemed necessary to continue the promotion of future global vehicle design efforts, ranging from standardizing side-impact dummies to tail-lamp lens colors.[2]

The relationships and interrelationships between parameters has never been easy to determine, due to the complexities involved. Some efforts have been made to identify the more straightforward cases, including the following example.

The increased costs associated with noncompliance with the safety standards could include fines levied by the government for noncompliance and recalls or internal company campaigns to correct defects on vehicles in the field. The manufacturers follow a self-certification format, requiring the government to prove that the company vehicles do not comply with the standards. The regulations require that a new vehicle, crashed frontally into a flat, immovable barrier at 30 mph at a minimum, shall not exceed the following fatal damage levels to front-seat driver and passenger anthropomorphic dummies: 1000 on the head-injury criteria (HIC) scale. The test speed has been increased to 35 mph, which has now become a new de facto standard adopted by the manufacturers. The National Highway Traffic Safety Administration (NHTSA) a unit of the U.S. Government Department of Transportation, contracts with independent laboratories to confirm

---

[2]Thomas T. Stallkamp, "Networking at the Boundaries," Conference Proceedings from the University of Michigan Automotive Management Briefing (Office for the Study of Automotive Transportation, Ann Arbor, Mich.) August 5–7, 1998, pp. 13–18.

the manufacturer's tests results. This is referred to as the New Car Assessment Program (NCAP).

Other tests require similar performance during side and rear impacts at nearly equivalent speeds, although the test conditions are different. (Side and rear tests involve the target vehicle either moving laterally at a slower speed than a moving barrier, or to be stationary, respectively). Other, higher-speed (40 mph), deformable offset barrier crash tests, which are in development by various research institutes, could become future requirements. Some manufacturers are already anticipating this outcome and are testing new designs for compliance.

Another dilemma posed by the safety regulators is the problem of multiple vehicle collisions, in particular between two vehicles where one vehicle is substantially heavier and stands higher than the other. Due to the laws of physics and the typically asymmetric payload characteristics of light-duty truck–derived vehicles, there is often a mismatch between the bumper systems of the former and small cars. This can negate many of the designed-in safety features of the smaller vehicle, if it is overridden by the larger vehicle. Fatalities and serious injuries occur more often from these impact circumstances than from single-vehicle collisions.[3,4,5]

Even when the standards did not originally apply, such as the case of standards designed for primarily for passenger cars, lack of crashworthiness by truck-derived multi-passenger vehicles such as vans and sport utilities can result in bodily-injury lawsuits and large jury awards for punitive damages, particularly when a defect is identified as a probable or definite cause of the injury or death.[6]

Meeting the standards can require literally hundreds of crash tests, for both developmental and final certification. Although an increasing amount of development work is being done through computer simulation, verification typically requires many crash or sled tests if early results are mixed. The major requirement now involves passive restraint system compliance, from both front and side impacts. Air bag systems will soon have to protect a wide range of occupant sizes and weights, from infants to full-sized

[3] Anita Lienert, "Sport utilities and cars face off," *Chicago Tribune*, sec. 12 (Transportation) April 19, 1998, pp. 1, 5, 6.

[4] Gregory L. White, "Car/Truck Crash Emphasis Should Be Side-Impact Protection, Producers Say," *Wall Street Journal*, May 26, 1998, p. B4.

[5] "NHTSA's Chief Says Cars, Light Trucks Need Design Changes," *Wall Street Journal*, June 2, 1998.

[6] Milo Geyelin, "Costly Verdict: Why One Jury Dealt a Big Blow to Chrysler in Minivan-Latch Case," *Wall Street Journal*, November 11, 1997, pp. A1, A12.

adults. This will probably involve several different systems for predicting who or what is occupying the front passenger seat. The level of sensor sophistication needed will have to differentiate between an occupant and an inanimate object, and to predict whether the occupant is properly positioned with respect to the primary restraint or belt system. In the past, manufacturers had to certify passive restraints for both belted and unbelted occupants, which resulted in, some say, overly aggressive bags to protect the heaviest unbelted dummy. This further resulted in serious injuries and deaths for some smaller adults and children, even though properly restrained. The proposed new regulations are being debated among safety advocates, regulators, and manufacturers as to how they will be applied and when.[7,8]

Vehicle manufacturers used to resist outside efforts to improve safety, but more recently they have begun marketing the value of safety system ratings published by the government and safety research institutes. Manufacturers frequently will phase in new compliance equipment earlier than originally planned or necessary to minimize the disruption to production lines and suppliers. Manufacturers usually do not dispute the government's claims of lack of compliance to an existing safety standard. However, in at least one case, a manufacturer fought back and had a fine and judgment overturned by a federal appeals court.

Priority, for safety testing, is usually provided inside the company so that the tests are passed with a sufficient margin for error, and to allow time for any last-minute changes that might be required. Vehicle mass, optional equipment installation rates, and the sales mix of vehicles can all affect how accurate the predictions for compliance will be. As vehicle mass increases—because of decisions to provide additional comfort, convenience or content—structural crush initiation time may decrease for absorbing front-end and side impacts and for activating passive restraints to protect occupants from jury. This could lead to higher-than-normal head-, femur-, and thigh-impact injury loads, resulting in more serious injuries. Thus, every effort is made to strive for minimum mass and, in some cases, also minimum manufacturing complexity, which can adversely affect quality durability and other parameters.

Ultimately, a new vehicle may start with a "design target" mass and cost that increases due to the addition of body or chassis structural support

[7] Anna Wilde Mathews, "U.S. Seeks New Air-Bag Requirements, but Car Makers Say Tests Are Flawed," *Wall Street Journal*, September 1, 1998, p. B4.

[8] Anna Wilde Mathews and Joseph B. White, "Car Makers Face New Rules to Make Kid Seats Safer," *Wall Street Journal*, November 6, 1998, pp. B1, B4.

that will permit the particular standard to be met. Then compensating mass and cost reductions may have to be found through material substitutions, reduced standard or optional equipment content and complexity, and so on.

A similar scenario typically exists for the emissions and fuel economy activities. Noncompliance penalties can be imposed on the company by the Environmental Protection Agency (EPA),[9] or additional taxes levied on vehicles with significantly higher-than-average fuel consumption. In certain times, the government has let the market determine its own incentives or disincentives, depending on fuel supply and demand. Vehicle curb mass plus certain optional equipment can affect where the vehicle fits into the "inertia weight categories" (which increase by 125 lb increments) used by the EPA to determine the chassis dynamometer setting for the emission tests. If the vehicle is not close to one of the weight limits, additional mass may not have a significant effect. However, if sufficient additional mass is added, and the weight category increases with resulting increased fuel consumption and emissions, the vehicle may have trouble passing the emission certification test and/or have higher-than-average fuel consumption. The costs to reduce mass can be enormous, requiring expensive material substitutions of aluminum or magnesium, to reduce the mass gain. Automobile manufacturers are loath to spend money to correct problems that might have been prevented early in the design phase, particularly for those that have no perceived benefit, or seem invisible to the customer.

There are also timing factors involved with these decisions. If everything planned works out, the problems disappear, but that outcome should never be relied upon as the first line of a backup strategy. Alternative plans that attempt to anticipate failures should be considered.

Parameters that more directly affect how a manufacturer conducts business for profitable growth are as important as the regulatory factors. The actual weighting of parameters will be determined by the individual project circumstances. Generally, internal manufacturing and quality improvements are driven not only by financial requirements but also by the market. Unfortunately, some of these latter issues conflict with the aforementioned regulatory parameters, in their relationship to mass. For example, cost of vehicle ownership considerations have also recently ascended the priority scale. Depreciation and insurance both rank higher than operating and finance, in the list of ownership factors that can affect vehicle purchase decisions. Depreciation is affected as much by resale value, as determined by regional and local used-vehicle auctions, as by the

---

[9]Natalie Hopkinson, "Honda, Ford to Pay Nearly \$25 Million in Emissions Case," *Wall Street Journal*, June 9, 1998, p. B11.

manufacturer's suggested retail price, leasing plans, less rebates and other discount policies. Warranty and/or defect recall history also plays a part in the depreciation calculation. Depreciation and recall information data is available from auction service companies and the National Highway Traffic Safety Administration (NHTSA), respectively.

Insurance costs are calculated by the insurance industry, based on their individual and collective collision and comprehensive loss experience, on each make, model, and year. Industry data is generally grouped by vehicle size but does not describe it by individual component or system. Collision estimating service bureaus do have some component-level repair cost data which can be purchased. Vehicle collision repair costs have a direct bearing on insurance costs. The other component of the physical damage coverage is comprehensive, which covers all other perils except collision. Theft typically represents approximately 40% of the comprehensive loss experience. Agencies such as the Highway Loss Data Institute in Arlington, Virginia, and The National Insurance Crime Bureau in Palos Hills, Illinois, compile and report vehicle theft statistics.

Improvements in both warranty/defects and repairability frequently, but not always, involve increases in mass and/or manufacturing complexity. Designing for damageability and repairability can mean moving vulnerable components out of primary load-path impact zones, typically located in the front corners of the engine compartment. It can also mean redesigning structural components to absorb more impact energy without allowing the damage to propagate further inboard. However, a potential conflict with another parameter exists when structures designed to crush for optimum occupant safety at typically higher speeds (15–35 mph barrier equivalent velocity [B.E.V.]), are redesigned to make them more crash-resistant at the typical accident range, (i.e., 8–12 mph B.E.V.). In most cases, unless a sacrificial configuration (i.e., a structure that collapses to save a more expensive component and is economically replaceable) is available for the low-speed scenario, high-speed occupant safety will usually take precedence over low-speed damageability.

The current debate over the effectiveness of passive restraints is concentrating on the most likely crash situations in which air bags can reduce injuries. Vehicle manufacturers need to know which mix of accident conditions, how many and what type of occupants, and what combination of restraint systems offers the best overall level of protection for the most people. Unfortunately, a purely statistical approach will probably not be completely acceptable to all concerned, due to the political content and quantity of nontechnical issues in the decision process. In spite of this dilemma, many manufacturers are installing expensive and, in some cases, questionably effective technology (e.g., side air bags), which may not help save occupants who are out of position or not belted, in the hope of gaining sufficient real-world experience without causing additional injuries or deaths in the process.[10]

Manufacturability can be analyzed by calculating the time required to manufacture the vehicle, including all of the component part costs, the labor to fabricate and assemble them, all finishing operations, paint, materials, and so on. Manufacturing complexity is defined as all of the operations necessary to fabricate and assemble the vehicle. Any extra operation required due to the addition of any new option, accessory, safety system, and similar, adds to the mass as well as complexity. Some design iterations may result in additional operations without adding mass, such as altering the shape of a structural component to trigger a different crash pulse. This action may not increase the mass, but it will undoubtedly change the complexity cost, particularly if new tooling is required.

There are other concepts involving manufacturing. One is "lean manufacturing," where components, systems, and vehicles are built according to actual sales orders instead of by trying to anticipate future sales. "Flexible manufacturing" has assembly and fabrication tooling that is flexible in operation so that two vehicles designed from different platforms can be built on the same assembly line. "Agile manufacturing" is more of an organizational model that facilitates adaptations in the production processes in response to ever changing market conditions.[11] Many, if not all, of these philosophies are geared to developing fixes for some of the problems faced in the assembly process, brought on either by poor designs or by vehicles with high complexity. All of them target wasteful practices that, through their elimination, save time in building the vehicle. They are essentially "after the fact" procedures intended to handle an ever-increasing list of option content loads and combinations of power-trains, trim levels, colors, and body styles.

Designing for manufacturing and assembly has a slightly different meaning. It was conceived to analyze how new vehicle designs could be simplified, through elimination of unnecessary fasteners, and by combining or eliminating components or operations. The efforts began with the standardization of fasteners and the elimination of low-volume trim colors and paint. Later, techniques for assembling components without fasteners,

---

[10]Gregory L. White, "Now Car Makers Rely on Air Bags for Side Impacts," *Wall Street Journal*, July 6, 1998, pp. A17, A22.

[11]Drew Winter, "Agile Manufacturing: Can It Really Make Big Operations Nimble?" *Ward's Auto World*, April 4, 1994, pp 40–41, 44.

by using snap-fit mating parts, welding and/or adhesives, were developed. Finally, software was developed to analyze the design while still in formulation, to find the simplest and most cost-effective way to assemble the component. It has also been utilized to design plant and equipment layouts.

Within the parametric analysis format, several other factors need to be considered in the assembly process. They include accessibility and disassembly for repair, and material separation for recyclability. While these requirements are not currently legally binding, future market or regulatory actions could change this situation. As was stated earlier, excessive repair costs can adversely affect insurance costs. Densely packaged underhood components can transmit low-speed impact damage that could also increase repair costs. However, some designs can accommodate both ease of assembly and disassembly/repair operations. Front and rear bumper reinforcement subassemblies, welded and bolted to chassis rails during assembly but needing only to be bolted together after repair, are current examples.

Design for manufacture and assembly has yet another meaning for suppliers, who are now being assigned more of the vehicle to not only manufacture but also design the components, subsystems, and systems for. In some cases, major suppliers are considered "system integrators," because they design, develop, and manufacture entire interior systems, including instrument panels, door and other trim panels, headliners, passive restraints, among other things. This type of supplier "owns" much more of the overall vehicle content that earlier vendors, who just furnished the parts as designed by the OEMs. Now, supplier input is considered vital to the success of a new product program. The supplier is more responsible for not only quality, price, and delivery, but also for process improvements, developing and maintaining partnerships with other tiers 1, 2, and 3 suppliers, and initiating additional new product innovations. Ideas such as modular construction, where various systems are combined into a subassembly for installation, have been well received and are cost-effective. However, consideration for the separate serviceability of functional modules such as bumpers, lamps and cooling system must be maintained. Moreover, modular construction is a volatile issue due to the potential conflict of nonunion workers from suppliers who are installing equipment on union-made vehicles.

With increased responsibilities come increased risks of problems, mistakes, and missed opportunities. Communication is a primary requirement, and it begins with including suppliers in a new program right from the beginning. Another requirement is to have common data interchanges—

computer systems that can "talk" to one another—and the trust that comes with sharing the design of a new product. "Computer-integrated manufacturing" (CAM) is one descriptive term that covers everything from common design software to tooling interfaces. Since most major suppliers work with more than one OEM, on simultaneous new product programs, the need for a common electronic data interchange is imperative. Several have had to install two or three different CAD/CAM systems and train their staffs to operate each one separately—an increasingly expensive situation.

Vehicle quality can be defined several ways, such as a lack or minimum of defects, a lack or minimum of customer complaints, high ratings in customer satisfaction surveys, and so forth. There are also several methods for insuring that quality has been designed into a new vehicle program, up front. Without digressing into specific or proprietary programs, it is still possible to outline the basic philosophy involved.

All of the various programs start with the need to identify and measure a baseline condition. If the current product is to be replaced or redesigned, documentation is necessary in order to determine how much improvement is necessary. Some programs seek to quantify a "loss function," a quadratic description of any component's or system's specifications that deviate from nominal settings or calibrations. These errors, quite often resulting from previously unidentified sources and referred to as "hidden losses," must be eliminated as part of any improvement process. Market research may help to gauge the scope and extent of these "hidden" problems in the eyes of customers. A technique known as "Design of Experiments" can be utilized to develop the inevitable interactions between some of the problems. Suppose for example, that, during a new model development, mass reduction has resulted in an accelerated durability problem of a suspension component, but a redesign has restored the durability but also affected the NVH (noise, vibration, and harshness), characteristics of the whole suspension system. "Design of Experiments" would permit the engineer to find the optimum mass to minimize the vibration and still satisfy the durability criterion, by utilizing a matrix analysis technique to sort through all of the experimental data calculations very quickly.[12]

Another parametric exercise is the ongoing quest to improve fuel economy, through a combination of mass reduction, power-train and driveline efficiency improvements, aerodynamic changes, and regenerative braking systems. All of these advance systems are theoretically capable of perform-

[12]W. R. Carey, "Tools for Today's Engineer" in *Strategy for Achieving Engineering Excellence*, SP-913 (Society of Automotive Engineers, 400 Commonwealth Dr., Warrendale, PA, 15096-0001 1992), pp. 46–52.

ing together to achieve an approximately 300% improvement in fuel economy for a current full-sized four-door sedan, without sacrificing safety, convenience, features, luggage capacity, or any other socially desirable benefits. The trade-offs between the technologies described would make the ideal case for a design of experiments, to find the optimum performance, sizing, and specifications for the vehicle and all of its systems, to meet the above goal.

Competitive products should also be included in the study, so that the product in question can be compared against them. Tearing down vehicles into their component parts is another method commonly utilized for bench marking or establishing "best in class" or best manufacturing practices.[13]

After analyzing the survey, focus group, or tear-down results, a further comparison should be made with internal company studies or warranty returns, to further identify potential quality issues that may not have been correlated to a design problem on the current product. This could include defects that were caught in the manufacturing or assembly plant environment, and corrected prior to shipment. Others could be assumed to have been random occurrences, such as computer-based codes erroneously displayed on an instrument panel or on a diagnostic tool during routine servicing but never traced or explained adequately.

From both body and chassis standpoints, NVH is often considered a quality parameter. Minimizing NVH should be addressed in the design phase, along with the fit and finish of exterior panels, consistent gaps between bolted closure components, squeaks, creaks, and vibrations felt through body components, since these are primary issues, both internally and to the customer. Data on all of these situations needs to be collected and analyzed as part of making corrections in the new design. The cost to correct them may add both mass and complexity, although not necessarily.

Performance parameters include adequate power for passing acceleration and towing capacity, braking, lateral stability for maneuverability, among others. Power-train sizing, optional equipment content, and vehicle mass are all factors affecting performance. The regulatory parameters also play an important part in determining actual performance levels. Increasing engine displacement to overcome a lack of performance may not be an option, if emissions and fuel economy standards are compromised. Manufacturers naturally take the viewpoint that this development is not in the best interests of their customers, although some have produced what are

[13]Drew Winter, "Dissection and Nitpicking Inc.," *Ward's Auto World*, December 1998, p. 113.

essentially "low-emission vehicles" or dual-fuel vehicles, under the definition promulgated by the Environmental Protection Agency (EPA).[14,15]

Similarly, braking performance may also be sacrificed on asymmetrically loaded vehicles, such as pickup trucks, unless a more sophisticated rear brake proportioning valve, rear-wheel or four-wheel antilock brake system is specified. Additional components or systems such as a turbocharger or supercharger would add both mass and complexity costs. However, the increase in total product cost could be minimized if the volume of the "charged" engine assembly was higher for the turbo- or super-charged models than for the naturally aspirated version. This phenomenon could result from economies of scale in the engine installation for wiring and vacuum lines at the assembly plant and also in the component suppliers' operations.

Cost of ownership is becoming a significant marketing parameter. Depreciation, insurance, repair, operation, and maintenance are the primary factors. Collectively, they can be an effective positive or negative factor in the purchase decision. Some are interrelated with other parameters, such as the relationship between vehicle quality and depreciation, which is primarily an ongoing analysis of the wholesale used-vehicle market and of lending institutions but is also strongly linked to customer quality audits, and dealer warranty claim histories. Insurance cost is related to damageability and repairability statistics for which reside within the insurance industry. Operating costs are related to fuel costs and availability, and maintainability factors are related to warranty experience and quality issues. Powertrain reliability is a direct contributor to maintenance costs.

Most of the databases on these parametric factors reside within the manufacturers or are available from key information providers. Commercial and/or rental fleets have extensive serviceability and maintenance records, which may be analyzed with their permission. Care must be exercised to insure that the proper vehicle data is being utilized. Model year changes must be documented accurately, to prevent mixing different vehicle types together or losing key information in the mix.

After analyzing the data, trade-offs or compromises may have to be developed, particularly in the case of conflicting data. For example, there could be a situation where a new design calls for the need to add mass, in order to solve an NVH problem, but results in an unacceptable risk in

[14]Joseph B. White and John J. Fialka, "Auto Makers Lobby Against a Proposal To Toughen Truck-Emission Standards," *Wall Street Journal,* November 4, 1998, p. A8.

[15]Joseph B. White, "New Clean-Air Rules for Light Trucks Set Battle Stage," *Wall Street Journal,* November 9, 1998, p. B4.

meeting a crash standard or the emission regulations. If no mass increase is permitted, changing the shape of the component may eliminate the vibration problem. However, manufacturing complexity may increase, with a resultant increase in fabrication cost, which could decrease the variable profit. These kinds of problems are not easily soluble, but the trade-off process can be more easily accomplished when all of the pertinent data is available and displayed in an up-to-date format.

The cost to improve vehicle quality, decrease damageability and repairability concerns, and increase durability and reliability seem to vary directly with mass and manufacturing complexity. Finally, improving recyclability may typically add mass and complexity, to account for the need to disassemble as well as assemble the product and to allow ease of separating different materials.

Within the parametric analysis, certain components might be targeted for minimum complexity cost increases while others might be targeted toward minimum mass considerations. More will be stated on this premise later in the text.

6.5 AEROSPACE INDUSTRY

Aircraft are even more dependent on mass than ground vehicles are. Mass determines nearly every parametric value involved in aircraft design and component selection. A list of typical customer requirements for an airliner is given below. (*Note*: Since this text is not specifically a product design manual, determination of the target values of the following parameters for a given aircraft design is considered beyond the scope of this analysis.) The following is a brief description of aircraft design and engineering processes:

After receiving the requirements from the customer (i.e., airline or government), initial layouts of the airframe are developed. Dimensions and specifications of engines, seats, and instruments and other equipment are requested and received from suppliers. Initial estimates of mass, aerodynamic drag, and lift characteristics from the wing dimensions and early wind tunnel tests are obtained from models and CAD studies. Other computer simulations are then performed to predict flight envelope calculations and stability criteria.

At the same time, manufacturing materials are undergoing environmental exposure tests at the company and suppliers' laboratories. Full-scale mockups of instrument panels, interior layouts, and galley and toilet configurations are being refined in the styling studios. Actually developed for initial concept testing, the refinements are based on customer review and approval. After a review of all test data and survey results, the final design is transformed into engineering and manufacturing drawings through the use of a digitizing machine.

Performance Characteristics

Aerodynamics (lift, drag, moments, etc.)

Altitude (maximum ceiling and economic cruising range)

Speed (mach number, maximum, economic cruising, stall, sink, takeoff and landing)

Range (miles or nautical miles, takeoff and landing distance)

Specific fuel consumption and capacity

Number of passengers and crew

Cargo capacity

Total gross weight (mass of aircraft, fuel, cargo, and passengers)

Aircraft empty weight

Crashworthiness and survivability

Durability and reliability

Repairability and maintainability

Serviceability (accessibility)

Selection and relative weighting of parametric values are based on market research involving negotiations with the principle airlines and/or secondary or "feeders," and with government authorities in the case of civilian and military aircraft. These groups of customers will determine the effective range of the parametrics and influence virtually every phase of the new product development process.

Civilian commercial passenger aircraft profitability is typically measured by the cost per seat-mile. The total cost is distributed between direct and indirect or overhead costs. Direct costs include all direct flight operational costs, such as fuel, crew salaries, maintenance, depreciation, and insurance of the aircraft. Indirect costs typically include depreciation of ground facilities and equipment, marketing expenses, administration, and general office overhead.

Revenue is the other determining factor for profitability, besides cost. Revenue is estimated on the type or class of seats and the number sold. The average coach fare, typically used in these estimates, is based approximately on the distance traveled. "Special" fares for excursions and discounted sales are based on the time remaining prior to the flight schedule. The number of seats sold divided by the total number available is the load factor. Revenue per seat-mile is the average (i.e., coach) fare times the load factor.

A breakeven analysis utilizing only the direct costs could be used to predict the basic economic value of a new aircraft design. The breakeven load factor is the minimum expense for a given flight. The total cost breakeven calculation must also consider all of the indirect costs, including the opportunity costs associated with how much additional gain could have been obtained if the development and manufacturing funds had been invested in some other venture. The discounted cash flow or net present value of those funds, compared with other opportunities, determines the economic success of the new aircraft. Depreciation, the discounted value of the aircraft purchase price, amortized annually for 10 years, is also an important factor is the above calculation. The economic decision should compare the net present value of the new aircraft to that of an expected normal rate of return on an investment of equal risk. If the net present values over the depreciable time schedule, added to the aircraft's salvage value, exceed the return expected from equivalent risks, then the investment in the new aircraft should be pursued further.[16]

Design for the new aircraft usually starts with an initial sizing exercise, utilizing either basic information from earlier aircraft configurations or industry data on engine performance specifications, the mass of standard components, published airfoil characteristics, and so on. Several of the more significant factors include the following:

Airfoil geometry and sizing (aspect ratio, wing sweep, taper ratio, twist, incidence, dihedral, vertical location, wingtips)

Airfoil lift and drag characteristics (design lift coefficient, stall, thickness ratio)

Reynolds number (air pressure and viscosity–related forces)

Control surface sizing and balance

Tail geometry (arrangement, volume coefficient, spin recovery)

Thrust/weight (power loading, horse power/weight, thrust matching)

Wing loading (stall speed, takeoff distance, landing distance, cruise, loitering, instantaneous and sustained turns, climb and glide, maximum ceiling)

Configuration layout and loft (wing/tail layout, wing location, wetted–area determination)

Structure

[16]Raymer, Daniel P., *Aircraft Design: A Conceptual Approach* (American Institute of Aeronautics and Astronautics, Washington, D.C., 1992), pp. 501–517.

It is assumed that the Simultaneous Engineering cross-functional teams will have at least one expert from each of the following areas:

Weights
Aerodynamics
Propulsion
Structures
Ground support/maintenance engineering
Manufacturing engineering
Avionics (aviation electronics)
Product planning
Cargo/passenger/weapons specialist (military)
Landing gear
Flight systems/stability and control/simulators
Cost planning

For military applications, the "envelope" (i.e., altitude, range, and speed), specified by the mission requirements, will determine the basic aircraft configuration. A fighter/interceptor may be required to reach its maximum altitude, with a maximum weapon load, within a certain time limit; remain at that altitude for another fixed time length; drop or not drop weapons against a target; retain sufficient fuel, and possess sufficient maneuverability, to gain or maintain superiority over an enemy aircraft; survive an attack from air or ground; return either directly or after fighting and evading, to a base. Attack and support aircraft may have a longer cruising-range requirement, at different altitudes, or more specific missions, including reconnaissance, electronic interference ground interdiction, cargo and passenger transport, search and rescue, among other requirements.

Besides weight or mass, other determinant factors that can affect the above parameters include the following:

Aerodynamic characteristics (i.e., lift, drag, and pitching moment coefficients)
Stability, stall, and spin tendencies
Takeoff and landing attitudes
Proximity to sonic conditions
Engine power output curves (thrust and thrust-specific fuel consumption at varying altitudes)

Control surface effectiveness

Instrumentation (control and display)

Communication equipment

The relationships between these factors are highly interactive but can lend themselves to Simultaneous Engineering analyses.[17]

6.6 HEAVY-DUTY OFF-ROAD/MILITARY VEHICLES

Specific customer groups will also have a great deal of influence in the design and engineering portion of this industry. Commercial utilization comprises almost all of these products, other than private motor coach recreation vehicles. Commercial requirements are no less important than consumer product needs. The principal requirements are for revenue and profit generation, while meeting the customer's demands.

Design and engineering operations for heavy-duty vehicles follow the same general format as light-duty and aerospace industries. Initial requirements from the customer are transformed into specifications by Product Planning. New product concepts are developed for customer approval. Dimensional data and specifications are received from major suppliers of power plants, transmissions, axles, wheels, tires, hydraulic systems, and implements. Component and system tests are performed in company and supplier laboratories on new materials and part configurations. Environmental tests such as temperature cycling, corrosion, and vibration are included in the lab work.

Based on test results and feedback from the customer, final designs and engineering calculations are prepared for customer approval. Final design details and engineering drawings for production are then completed and approved.

Some of the more significant design and engineering parameters are listed below:

Functional characteristics

Power plant and conversion

Track/wheel drive system

Payload type and capacity

Operational speed and range (with and without payload or equipment)

[17] Alfredo Herrera, The Boeing Company, personal conversation, June 4, 1999.

Operator safety and ergonomics

Instrumentation (control and display)

Stability requirements (center of gravity, with and without payload/equipment, outriggers, etc.)

Maneuverability (turning radius, approach and retreat angles)

Maintainability (serviceability, repairability)

Without further defining the specific outcome required by customers, there are a few general requirements that can be provided. Modular or building-block types of designs are prevalent in the following markets:

Soil, vegetation, and rock extraction and transport

Liquid cargo transport

Cranes and other high-lift operations

Firefighting and rescue

Medical transportation and injury evacuation

Refuse and recycling material removal and recovery

Passenger motor coach and recreational vehicles

Articulated tractor and trailer(s)

Common operational characteristics, for any given-size vehicle, in all of the above markets, include maximum payload, operating speed range, distance, curb or unloaded mass, fuel consumption and capacity, lubrication oil capacity, hydraulic system capacity, auxiliary power capability, towing capacity, among others. In the case of some equipment, mass is not a parameter to be minimized, as it can dramatically affect vehicle stability and handling, particularly in off-road situations. Typically, optional equipment does not present as many problems as do consumer goods, since commercial application specifications are controlled by institutional and/or corporate directives on cost. Thus, increase in mass and manufacturing complexity may be due more to "envelope stretching" than to adding optional features. (For example, heavier payloads may necessitate larger output shafts and bearings, which may in turn require larger castings.)

The market competitive position provides many of the current design constraints, although there are industry standards and recommended practices that tend to provide a form of unofficial regulation on some issues. A certain level of uniformity exists in the areas of safety equipment, instrumentation displays, color-coded maintenance appliances, operational control locations, and so on. Additional uniformity in other ergonomic issues could be designed in if the market demanded it. At the same time, new equipment entries from overseas create additional opportunities and market pressures.

On wheeled as well as tracked vehicles, axle torque application and traction as determined by track widths and/or shape, tire width, tread design, ballast, mean ground and inflation pressures can be important factors in how well the vehicle propels itself or tows other payloads in mud, rock, sand, or wet, slippery soil conditions.

Static and dynamic stability can also be important design factors. Center-of-gravity location, both fore–aft and vertical, will determine tipping angles (i.e., maximum safe operating conditions). Some agricultural tractor and implement combinations are sensitive to dynamic load pulsations, or porpoising, which not only can adversely affect the worked area, but can also contribute to operator fatigue and loss of concentration. These operating conditions should be designed to be, if possible, outside of the typical operational bandwidth of speeds and payloads.

Ergonomics and the total number of work tasks required by the operator can also be a factor in determining safe conditions. Monitoring of direction, speed, terrain angle, tool/implement orientation, potential obstacles or payload shift, and so on, can place an undue burden on the operator. New designs should consider eliminating as many unnecessary monitoring functions as possible. Audible signals, tuned to frequencies not attenuated by the frequency of the power plant, could be an alternative.

6.7   ELECTRONICS

Design parameters for large-scale integrated circuits can be listed into several categories. Size, geometry, functional density, event execution speed, power dissipation, and parasitic effects are all critical design factors. For microprocessors and other computer-based applications, new manufacturing techniques have been paramount for the last 15 years, as miniaturization and functional proliferation efforts resulted in high-density packaging and increased value in each unit. Some of the challenges facing integrated circuit designers and engineers include some of the same affecting any other new product: cost, included features, increased competition, and so on. In addition, further miniaturization, increased computational speed, greater reliability, compatibility with other systems, and a foreshortening product life cycle continue to exert enormous pressure on many market segments of this industry. Since products and markets are so diverse, this text will concentrate on a strategy for collecting and considering all of the essential elements inherent in a generic IC design, one that could be applicable to virtually any electronic product.

The merits and capabilities of solid-state semiconducting materials were discovered, developed, and commercialized within the relatively short span of 50 years. Primarily driven by the requirements of the aerospace industry, the constantly increasing payloads of electronics demanded that miniaturization, reduced power consumption, and increased reliability of onboard navigation, weapons control, and communication equipment be pursued.

While this text is not a design manual for ICs, a moderate amount of theory may be helpful in explaining how the interactions between key design parameters should be considered. The most basic semiconductor device, even in large-scale integrated circuits, is the a bipolar junction transistor (BJT), which typically contains three elements: an emitter, a collector, and a base. Controlling electrical current flow between these elements is the fundamental objective. Another basic semiconductor device is the metal oxide semiconductor field-effect transistor MOSFET, primarily used for voltage control in computer memory circuits, where internal and external capacitors are alternately charged and discharged, signifying memory storage conditions, with voltage signals.

Depending on the external circuit requirements, the device may be designed to perform as a switch, an oscillator, an amplifier, a rectifier or diode, a filter or capacitor, or as space charge storage. The methods employed to deliver the desired functions may depend on other engineering trade-offs, such as manufacturing cost, package size, operational speed constraints, power consumption, and so forth. Whether other devices such as resistors and capacitors are built in, or purchased and installed as separate components, may depend heavily on the above parameters.

Between the three above elements, which are connected to conducting paths attached to or embedded in a wafer of silicon, are other materials, deposited by various means, that provide for "gated" (highly controlled) current flow, only under specific, planned circumstances. These materials, normally also semiconductors, have crystalline structures with either excess electrons or "holes" (spaces) to store other free electrons. Materials with the most excess electrons are classified as metals in the periodic table. However, some metals are very conductive, resulting in current control difficulties. Some others, such as phosphorus, have fewer free electrons, but good donors nonetheless are known as *n*-type, or negative, materials. Acceptors, on the other hand, come from a different group of elements, of which boron is prominent, and are known as *p*, or positive types. *P* and *n* materials can be altered chemically through the inclusion of other semiconducting materials, such as arsenic, gallium, germanium, and others. This is called "doping," and is performed through the utilization of different processes including layering through diffusion, ion implantation, oxidation, chemical vacuum deposition, and etching. The positive and negative tendencies of the materials can be enhanced with the proper mix of impurities, thus furthering their capabilities.

The amount and/or rate of charge and discharge accumulation and current flow is a function of the type and amount of impurities that are intentionally mixed into the fabrication process. These materials are deposited in relatively thin or thick uniform layers. The proximity and thickness of differently "charged" materials can determine how well they conduct current flows; i.e., how much, how fast, and what parasitic effects are also produced? For very large scale devices, containing thousands of transistors, the interconnections or channels need to be as short as possible to maximize transmission speed and minimize capacitive tendencies between adjacent channels. In addition, if the layers are too thin in spots, "punch through," similar to a short circuit, can result, which tends to drain away an additional amount of the space charge, preventing or severely restricting operation of the device, which is the charging and discharging of a load capacitor. Separating space charge layers with an insulating silicon oxide layer was an early usual control method. One drawback inherent in silicon oxide is the potential leakage from porosity. Sophisticated manufacturing techniques are available to insure high-quality oxide layer thickness.

The perspective of the IC design engineer will need to consider the following tradeoffs, within the requirements specified by market conditions:

Packaging size may be constrained by the minimum channel width of adjacent conducting path (breakdown voltage).

Maximum speed to charge and discharge a given circuit capacitor.

Number of integrated vs. "wired" components (capacitors, resistors, etc.).

Number of external connections.

Device scaling (reduction in size, i.e., number of devices per square centimeter).

Size of heat sink (power dissipation factor).

Another decision has to do with the operational characteristics desired in the IC. Two of the previously mentioned devices are bipolar junction transistors (BJTs) and metal oxide semiconductor field-effect transistors (MOSFETs). BJTs are mainly current control devices, while MOSFETs are primarily voltage control devices, most commonly utilized in computer random-access memory banks. BJTs are more effective as nonsaturated analog devices featuring high gain and low noise. MOSFETs, on the other hand, usually operate in two modes: grounded and saturated (i.e., 0.8 volts and 2.4 volts, respectively). In a digital circuit, 0 is ground and 1 is saturated. The binary numbering system utilizes only 1's and 0's, which is complicated for manual operations but ideal for processing digital computations and transmitting the information quickly. Such operations can be utilized to activate a switch, engage an inverter, flip-flop circuits, charge and discharge capacitors rapidly, and so on.

Another analysis that needs to be performed is the system in which an IC is to be installed. Assuming that a generic IC is to part of a larger generic system, designed, as a transfer function, to provide a number of outputs based on a number of different inputs, some of the major design parameters are listed below:

System architecture (hardware and software)

Circuit complexity (chip size, minimum scaling geometries [limited by processses], data volume limitations)

Power dissipation requirements

Functional characteristics (What is the device supposed to do?)

Input and output number and type (analog, digital)

Operational environment

Diagnostic codes and procedures

Reliability (system, component)

Safety

Obsolescence (replaceability or upgradability potential)

When the systems engineer has gathered all of the above information that is available, discussions with potential suppliers can take place. One of the first decisions to be made is whether standard hardware devices can fulfill all of the requirements. Even if a standard IC is available with upgradable features, a custom-designed IC may be less costly from a system standpoint than the standard component approach. It depends on the system complexity and how well the end user will be able to utilize the outputs. Physical package limitations may preclude locating all of the subsequent upgraded modules in the same space.

The iterations between a standard integrated circuit and a completely custom design consist of the following:

Gate Array: A VLSI chip containing a large number of identical transistors, arranged to perform cascading or other logic functions. As the circuit complexity increases, interconnections between the arrayed elements become increasingly difficult, with long metal conducting paths and potentially serious parasitic conditions. The limitations of wiring dictate the maximum level of circuit complexity achievable.

Master Slice: This scheme has groups of identical transistor arrays that are interconnected at common boundaries. Interconnections can become more unique than gate arrays, however, circuit complexities are governed by the limitations of the background cells or the clustered devices.

Master Image: An improved version of the master slice whereby the background levels can be customized to best fit the circuit design. Every chip is unique, but in fitting them together in an optimized fashion, the same functions are connected the same way on each physical layout. Wiring channels and interconnections are the constraints for this design practice.

Hand-Honed Macros: This is where the circuits are designed to fit together, optimized for space and function. The entire circuit is a module that has to be designed in its entirety, in order to be functional.

The design options described above each require decisions related to the trade-offs among circuit feasibility, logic complexity, design time and cost, performance, chip size, turnaround time to market, and the skills and resources available to the firm.[18] The decision process involved with circuit design begins with how much of a standard component can be utilized in the circuit. If less than approximately 10% of the component function is usable, a smaller, denser chip can probably be designed faster and manufactured cheaper than a standard model can be incorporated. If more than 50% of the standard chip can be utilized in the circuit design, a custom-designed chip will probably take longer and cost more than incorporating the standard model. Between 10% and 50%, other parameters will need to be evaluated.

Circuit density is a major power or heat dissipation factor that may limit the total number of circuits. Other limiting factors include the following:

Transient currents (which can be minimized)

Net capacitances (i.e., connector lengths, diffusion areas, gate widths)

[18]*Design Methodologies, Advances in CAD for VLSI,* vol. 6, S. Goto, ed. (Elsevier Science Publishers, Amsterdam 1986), p. 19.

Controlling circuit energy (through lower power supply voltage)

Long cycle time to maximize performance

All of these factors create an interdependence between performance, density, and power that must be resolved in light of maximizing the output of the circuit. This dilemma is typically solved in a custom design.

The major cost parameters can now be presented for individual devices. The results can be multiplied by the total number of devices to obtain the total cost.

Product design and development cost

Wafer production cost

Semiconductor processing costs

Bonding and interconnection process

Yield testing and evaluation cost

Packaging cost (plastic injection molding, etc.)

Assuming that either a standard IC will fit the requirements or a supplier will furnish a custom-designed equivalent, the rest of the steps in the design sequence are as follows:

Select the necessary input sampling rate and amplitude range (typically twice the highest signal frequency, dynamic range vs. noise level, etc.).

Initiate algorithm development (rate of functionality implementation in computations per second).

Begin hardware and software development.

Consider ground support equipment requirements.

Begin development of end-user interfaces.

The above list lends itself very well to a Simultaneous Engineering format. Discussions with various specialized engineering groups both inside and outside the firm will add substance to the outline thus presented.[19]

6.8 CONSUMER PACKAGED GOODS

Design of consumer packaged goods is beyond the scope of this text due to the variety, style, function, size, and shape of the products. Chapter 15

[19]William B. Ribbens, Electronics consultant, personal conversation, March 28, 1999.

contains guidelines for packaging design that may be helpful for product design as well.

6.9   AJAX MANUFACTURING COMPANY

One of the key problems Ajax engineers faced was designing the LVDT mechanism so that the manufacturing department could hold the tolerances for proper instrument operation. This involved specifying machining tolerances as low as ± 0.0000010 in. or ± 0.000025 mm. Ajax's current production equipment was not capable of consistently delivering this level of quality. Therefore, new lathes, milling machines, and centerless grinders were ordered for the new M-2000 device. The new machines were all computer controlled, and the operator would be there to monitor progress and make sure everything worked as planned. Production engineers were to be trained in how to program the machines to utilize the dimensions from the CAD/CAM design software, to feed them directly into the lathe, which would then perform the required cuts.

Material selection was another problem that was solved by the engineering program, which matched the requirements for stability over the temperature range, for good machinability, and for easy anodizing for corrosion resistance. A 6000 series aluminum alloy was chosen for the case and moving components, except for the steel pole piece. The pivot bar was to be fabricated out of a similar aluminum alloy, with the diamond stylus adhesive bonded in a hole in one end. The instrumentation was to be packaged in a separate module that could be positioned conveniently on its own pedestal or clamped to an arm or other safe and proper location. The output screen was configured as an vacuum fluorescent screen with the option of patching the output directly to a color monitor, through standard serial port connectors. The circuitry was to permit measuring virtually all known surface roughness parameters, along with roundness, waviness, and profile. The software even allowed for future parameter development, if required, when the unit was sent back to the plant for recalibration.

7

PROTOTYPE DEVELOPMENT

7.1 INTRODUCTION: ASSURING THAT THE NEW PRODUCT DESIGN COMPLIES WITH ALL TEST STANDARDS

This phase of the new product development process involves validating that the design meets all of the product objectives specified in the product planning document and in turn, carried out in the design and engineering phase. The objectives of prototype development are to assure that the design can be built as designed and will also meet all customer requirements.

Most industries and major firms have their own internal certification or acceptance standards for product performance. For the five industrial groups under analysis, the following is a partial list:

Department of Transportation, National Highway Traffic Safety Administration (DOT/NHTSA) federal motor vehicle safety standards on passenger cars

EPA certification for emissions and fuel economy, for light-duty vehicles

DOT Federal Aviation Administration (FAA) certification of air worthiness and crash worthiness requirements, for airliners

U.S. government military specifications and military standards, for acceptance of militar, ، uipment

Underwriters Laboratories, (UL), for approval of electrical equipment

Federal Food and Drug Administration (FDA), for approval of certain prescribed and packaged drug products.

7.2   PROTOTYPE TESTING AND EVALUATION

This particular topic will analyze testing of automotive, aircraft, heavy-duty, and electronic products. Consumer packaged goods is not included due to the variety and complexity of consumer products. A brief description of the testing of packaging is contained in section 7.2.5.

7.2.1   Automotive

This step involves extensive accelerated durability and environmental testing, with and without other system components, including destructive tests, and so on. In the past, the process frequently proceeded over the entire product design and development cycle, with sequentially built phases of prototype variants, each one closer to the final production configuration than the previous iteration. Test results from the earliest models were utilized to update succeeding generations, pending evaluation and subsequent component, system redesign, and new part fabrication.

A newer technique has, in some cases, replaced testing hand-built prototypes. This process involves computer simulation of stress fatigue and finite-element and rigid-body analyses to optimize strength levels of structural components and the entire vehicle. The analysis usually begins with component and subsystems, then considers body and chassis bending and torsion, followed by component durability and fatigue, and ending with full vehicle accelerated durability. However, simultaneous consideration is most often preferred so as to address any applicable trade-offs, up front. Design optimization can also be performed at this stage, with minimum mass targets, as an example.

Within component/subsystems analyses, joint stiffness of both unitized and frame vehicle bodies can have a dramatic effect on overall vehicle durability. These include roof, floor, and side pillars, front-end structure, underbody anchorages, and internal and external reinforcements. Analysis of closures (i.e., doors, hoods, decklids, and so forth). is also important for optimizing load and sealing performance. Other performance parameters include vibration, torsional rigidity and energy absorption during crash. These last two are potentially at odds for a hood, which needs to withstand flutter while folding properly during a front barrier test to limit windshield intrusion. Limiting door intrusion during side impacts requires a major design concentration, due to the interaction of the floor, reinforcements, pillars, hinges, and locks. Viewed as a system, finding the weakest link in the progressive chain of dynamic collapse modes can be modeled to provide satisfactory results.

Analysis of complete body and chassis bending and torsional rigidity is utilized to identify any durability and noise, vibration, and harshness (NVH) problems. Durability/fatigue analysis is used to model stress/fatigue on specific or localized vehicle areas, patterned to reflect the stress concentrations expected, once those areas have been identified. Previous model history can be beneficial in those cases. Finite-element analyses are utilized to represent relatively rigid (i.e., low displacement) elements such as fasteners, along with less rigid body structural elements prior to any stress analysis, since displacements are considered more accurate than stresses.

After the finite-element method has isolated the areas of the body structure needing further fatigue analysis, the model is subjected to input from road load data. The resulting fatigue history is compared to earlier configurations for improvement. Modifications are made, if necessary, and the model is rerun.

Next, full body/chassis fatigue is executed in the same fashion. NVH control typically involves the following three areas:

Controlling the magnitude of the source force

Controlling amplification/attenuation along various paths

Controlling the response at the occupant contact point

Since most designs have already been optimized for packaging constraints, minimum mass, and minimum cost to produce the minimum force, most NVH optimization will have to be performed within the confines of the second and third steps.

Both external and internal forces occur in the vehicle. External sources include road surface contacts through the tires, wind noise, and aerodynamic loading of exterior panels and glass. Internal sources include the following:

Power plant combustion reactions including the exhaust system

Power plant to power-train unbalance

Fuel supply slosh

Fan and compressor noise

Tire/wheel unbalance

Brake torque variation

Road surface irregularities

Aerodynamic forces

Some of the above effects can be eliminated through component or subsystem redesign, such as reducing the vibration tendencies of fans and compressors, by changing the metal gauge or shape of the blades or the mounting brackets on the air-conditioning compressor. More significant improvements require computerized analysis for optimization. Examples include minimizing seat sound pressure from single-multiple tire jounce inputs.

In early underhood packaging studies, a technology known as stereolithography (rapid prototyping) has been utilized to save time in obtaining prototype parts. The process works by a PC-run laser cutting through a liquid bed of a photopolymer, curing it as a flat planar surface. As each slice is cured, it is automatically lowered to make room for the next layer. Over time, the part is built up until completion. Shape is very close to true, but the part is nonfunctional. Round, curved, hollow, and partial or cutaway models can be constructed in the same manner, although complex shapes may require a final cure from an application of heat or ultraviolet light.

Other techniques include laser sintering, photochemical machining, and laminated object manufacturing. Laser object sintering is somewhat similar to stereolithography except a infrared laser beams are utilized to not only cure a flat surface, from a fluidized bed of photopolymer powder, but also to cut away portions of it in consecutive steps. Photochemical machining utilizes two laser beams to mold and cut simultaneously. Laminated object manufacturing starts with solid thin sheets of paper, plastic, or foil, and cuts shapes out for later lamination with adhesives or welding.

7.2.2 Aerospace

Full-scale prototype testing of aircraft has decreased as the science of computer-aided design and computer-aided engineering (CAD/CAE) systems has been perfected and implemented. Wing-stress distributions; interference drag, and thermodynamic effects of engine nacelles; flight stability derivatives; and other information necessary for the assurance of flight safety and economy are routinely calculated utilizing computer systems. However, manufacturers still perform many tests on the first few completed aircraft. Confirmation of stall speeds, spin characteristics, and recovery techniques and performance calculations are still necessary. This information is typically collected on flight test prototypes. In some cases, the first production of a new commercial airliner becomes the flight test model. In other cases, such as for military models, units are purposefully built for flight testing alone.

For specific flight test models, hand-built units are assembled utilizing prototype parts either fabricated from temporary tooling or through the techniques mentioned above, including stereolithography.

7.2.3   Heavy-Duty Off-Road Equipment

Construction equipment prototypes are early preproduction units that are tested extensively to confirm the capabilities designed into them for maximum safe loading, static and dynamic stability, surface compacting under tread, and intrasite movement limitations. Any failures or defects would also be noted and corrected in the regular production units. If problems occur in meeting the requirements of the above tests, this information would be important for potential advertising literature or technical specifications.

7.2.4   Electronics

The "breadboard" (i.e., hardwired) models, mentioned earlier, serve as pre-prototypes, providing feedback on problems and both favorable and unfavorable comments from customers. These inputs are utilized to tune or adjust the design and production facilities, prior to actual production. Early production units are tested extensively to uncover any other problems, as yet unidentified. Tests include vibration, temperature cycling, thermal shock, drop testing, and high-voltage spikes. Any failures would be corrected and retested before regular production commences.

7.2.5   Packaging Testing

Testing of the packaging would include the same environmental tests already described, in addition to destructive tests for tamper resistance. This would include penetration by sharp instruments and hypodermic needles, crushing from heavy objects, and exposure to ultraviolet light and infrared radiation.

7.3   AJAX MANUFACTURING COMPANY

The initial M-2000 design has now been completed for about two weeks. The machine shop has begun to fabricate some of the detail components. One of the more complex parts was going to be the case. Originally, this component was planned to be an aluminum extrusion with access for assembly and disassembly through each open end, then capped off with bolted access plates. However, it was found that in order to change the battery pack, the entire probe rod assembly would need to be removed first, which was not a cost-effective operation. A different location or attachment was needed for the battery pack. Finally, after a mini-brainstorming session involving most of Ajax's designers and engineers, it was decided to mount the battery pack, piggyback-style, to the outside of the case, on the side opposite where the probe rod exits the case.

The first 10 units were assembled by hand and carried to key customers for evaluation on the shop floor. After two weeks of production exposure, a number of operational problems were recorded and analyzed. The most prevalent was instability of the electronics circuitry, where drifting (i.e., required frequent calibration), voltage spikes, and power interruptions, plagued the early units. Noise emanating from the pilot gear train began to cause further problems with backlash and binding of the pilot rod against the case. The most encouraging portion of the evaluation was the consistency of the readings, based on comparison with the standard "surfaces," furnished by Ajax.

The electrical problems were diagnosed as defects in the circuit design, which was subsequently corrected. Some of the plated-through hole connections had a missing etch mask, which was overlooked in the first build sequence. The gear train noise and binding problems were caused by an improper gear material and incorrect backlash settings.

Several customers raised questions on the instability of readings when measuring a narrow bearing surface or O-ring seat. Apparently, it was difficult to hold the case, even when it was reshaped to fit hands better than in the early units, up against a small surface. One Ajax engineer had an idea for fixturing small parts and rotating them against a rigidly mounted pilot and tracer. After three weeks, a crude model was fabricated and demonstrated to the customer, who was enthusiastic about it. Other customers also tried it out and liked it, and the orders started coming in, not only for the modified rotary model but also for the regular M-2000. Having influential customers as part of the new product development process seemed to be paying off.

One final observation was made regarding the display screen, which had limited visibility unless viewing it straight ahead. One of the suppliers suggested mounting the display module on a swivel base so that it could be positioned for viewing through out a 360° circle.

8

MANUFACTURING

8.1 INTRODUCTION: VERIFYING THAT ALL MANUFACTURING CRITERIA CAN BE MET

Manufacturing is concerned with everything necessary to produce a product. This includes all of the processes heretofore explored, as well as those affected beyond, such as sales, service, recycling, and so on. Any product, by definition, has to be producible; however, there are degrees of ease or difficulty of manufacturability. These are caused by the difficulties of replicating the design, the effectiveness of the prototyping process to uncover problems, and the efficiencies of the manufacturing process. A producible design considers all of the factors that can affect how component parts are fabricated and assembled, how subsystems are fitted together, and overall product performance. Designing for manufacturability was addressed in the last chapter. This chapter is primarily devoted to those portions of the manufacturing process that can either positively or adversely influence the total product cost structure. Since virtually everything on the plant level can affect product cost in one way or another, this chapter will concentrate more on fixed or indirect costs than on variable or direct costs. In many instances, both types of costs are tightly intertwined, or are at least highly interactive.

All products have a degree of functionality. How well products perform their intended functions depends, to a large extent, on how close the production version is to the original design and/or final design intent, based on the development experience with a prototype. The potential accuracy of the design is also important. Manufacturing is usually based on a sequential occurrence of events, specified so that previous steps will provide a robust foundation and path for succeeding stages. However, some designs do not offer the above advantages, and may be so difficult to manufacture as to render them unfeasible. While it is unfortunate when this event occurs in the design stage, it is catastrophic when it happens in the manufacturing phase, after significant time and money have already been expended. More often than not, major and minor problems occur in the fabrication or assembly stages, as a result of design interferences, misaligned mating surfaces, component placement or packaging, or the build sequence. These problems can be corrected through redesign of the product and/or the tooling, to eliminate the interferences and misalignments, or by changing the build sequence, with the attendant increase in production costs and delays in getting the product to market. This is why it is recommended that prototypes be assembled utilizing hard or final tooling, so that the fit-up problems from temporary tooled parts can be avoided.

Product complexity occurs when variations in style, optional accessories, equipment, and other features, are continuously added to the original product in response to market conditions. It happens most often after the product has been on the market for several years and may need freshening, in order to boost demand. It can also occur anytime an original design is modified to add operations to components and systems, to provide upgraded value from that of the previous design. When additional or more complicated tooling for stamping, welding, assembly, or painting steps calls for more labor time, material, or parts for the product, complexity is increased. However, when parts are reduced by combining them with other parts, or when fewer operations are necessary than in an earlier version, complexity is reduced, but tooling costs are still increased because tooling has to change with the part. Component reductions can also have a major impact on repair, warranty, and recycling processes and should be considered carefully before making the decision.[1]

Is it possible to avoid the above situation? In the past, some product redesign was considered inevitable. Design, engineering and manufacturing were separate sequential operations, performed by separate departments with very little interaction, until the problems forced the different groups to cooperate in order to solve their mutual problem. When competition was limited or restricted to a few similarly sized firms, any additional product development costs could be readily passed on to consumers. If other competitors appeared on the scene, with new design methods, more efficient production, less entrenched management philosophies, and who were more responsive to customers, their products would be received readily if not enthusiastically. This resulted in enormous pressure among the original

[1] Mary Connelly, "Ford Cuts Trim Levels, Options to Simplify Assembly," *Automotive News*, September 28, 1998, p. 10.

market participants to regain competitiveness. The subsequent drive to do so has included reeducation of workers, new CAD systems, and production equipment. It also has entailed a new philosophy based on simplifying the entire design process, by focusing on specific parameters such as designing for manufacture and assembly. This system concentrates on all of the aspects of manufacturing affected by design, including fabrication of components, assembly sequence, and the costs associated with these operations, which can be readily analyzed for improvement. Other cost factors, including tooling, can also be analyzed with similar software packages that consider warranty, service, and recyclability issues.

Four of the five industry groups are now analyzed for their specific manufacturing requirements. Packaged goods manufacturing concentrates only on specific packaging operations.

8.2 AUTOMOTIVE

The automotive industry experienced one of the fastest growth cycles in history from just before the start of the 20th century until the Great Depression of the 1930s. Worldwide annual auto manufacturing reached millions of units within the first 20 years. Manufacturing methods moved quickly from horseless carriage conversions to integrated manufacturing, where raw materials entered one end of a plant, and finished vehicles were driven out the other end. Chain-driven assembly lines moved partially completed vehicles along which workers repetitively installed the same items on each vehicle passing their station. Workers were considered a unit of production, the same as a machine or the components being assembled. Production was continuously increased with higher conveyor speeds, more workers, and machinery. However, monotonous work contributed to fatigue, which in turn contributed to mistakes.

Production defects have led to customer complaints, recalls, and increasing warranty expenses. Automobile manufacturers have been consciously striving to reduce these expenses, through the adoption of new quality assurance procedures, which typically involve reviewing and renewing every aspect of the production process, including those of suppliers. With suppliers now shouldering much of the overall design responsibility, these efforts are accelerating. No detail is too small for scrutiny, as cost reduction and quality improvement are twin goals with equal priority. Highly automated plants, having been found too costly and lacking in consistency and accuracy, have been supplanted with additional inspection, measurement, and monitoring of incoming material, parts, and components.

The major improvement in manufacturability came from involving manufacturing engineers and workers in the vehicle design process. This happened not just on the periphery but right from the concept discussions, but also in every other design and engineering process. The same production personnel are often involved and responsible for fabricating prototype parts and designing and building experimental production equipment and tooling. The process is continuous as new design data is completed, successive prototype parts are tested, and the results are converted into hard tooling for stamping dies and welding fixtures. Pilot models are built utilizing production equipment but at slower line speeds, to try out the dies, to test mating part fit-up and gaps, and to evaluate components from suppliers.

After regular production begins, quality audits continue, throughout the model year, in order to monitor key parameters that make up customer satisfaction measurements by market research firms. Warranty claims from dealers are also studied, as the entire design team seeks to implement continuous improvement.

Key manufacturability cost parameters include the following:

Direct Manufacturing Costs

Purchased components

Fabricated components

Assembly labor

Finishing labor and material

Indirect Costs

Overhead

Safety

Regulatory compliance (in-plant)

Inventory

Quality (first-run launch and continuous improvement)

Recycling

Customer responsiveness and satisfaction

Inspection procedures

Production control information technology and product data management

While the above items comprise the majority of costs in manufacturing, other factors are also contributors. These include time to pick up and move components from one location to another; part mobility due to mass, shape, or size; lack of adequate space to install a part due to another component overlapping the same area; lack of accessibility to utilize standard hand or power tools for assembly; insertion of wiring, hoses, or tubes in long, narrow, or confined spaces.

Machining tolerances on components can cause problems both for being too tight and too loose, compared to that of mating components. Mismatched tolerances due to process incompatibility can cause fit problems as well as component performance degradation and premature failure. Heavy, awkward, or unwieldly subassemblies, such as seats, windshields, doors, and so on, can become a worker safety issue in addition to a productivity issue. Robots and other factory automation have helped reduce worker stress levels by taking over some of the more awkward and dangerous jobs. The same analysis is being utilized on fixtures and other tooling.

Total combinations of end items, including all color and size variations of components such as interior trim, moldings, seat fabrics and so on, can affect how well the assembly plant can handle their integration. Designing parts to fit together only one way is one method, useful even when right- and left-handed parts are packaged as a set. Designing and marking the shipping container to fit the parts being shipped is another method to control that the correct component will get to the right point to the line. Snap-fit parts are still another method for optimizing manufacturability. This method eliminates potential fasteners and adhesives that could slow production rates, by requiring waiting time for adequate tightening or curing. Recyclability could also be affected if disassembly is more complicated; i.e., cutting or breaking adhesive bond lines is cost prohibitive and potentially risky for worker safety. Removal of fasteners subject to corrosion may also require cutting with an acetylene torch. Elimination of fasteners also saves mass, which could become the most significant effect.

Many of the tasks planned for current and future manufacturing technology are being driven by the demands for improved product quality, customer satisfaction, and unit profitability. After the more obvious fixes have been implemented, the cost to improve quality through changing processes or creating new ones has to be analyzed in light of the other parameters, such as profitability. More than likely, new processes will be incorporated into new model vehicle programs, where planning strategies could be utilized to balance the needs of manufacturing costs as well as other design parameters.

Some of the new technology already introduced into the production processes include the following:

Body laser-welded tailored blanks

Hydroformed structural components

Flexible programmable fixturing

Intelligent sensors (operations utilizing fuzzy logic)

Computerized numerical control

Device-level communications standard

Single-step body assembly

A brief description of some of these technologies is as follows:

Tailored blanks: Two or more different thicknesses of aluminum or steel are laser welded, as flat stock, from separate coils. The minimal heat generated by the laser results in less warping or burn-through than conventional welding practices. The resulting blanks are then trimmed to a particular shape for subsequent stamping into chassis rails, pillars, or other complicated body parts, using a sequence of progressive dies.

Hydro-forming: Round tubing is filled with water, along with a small amount of oil for corrosion protection and sealing. The tube is then forced through a set of dies, which bend the tubing in different directions, while the water pressure inside keeps the tube wall from collapsing. Finally, the wall is pierced for mounting holes, and so forth. The technology is used to fabricate long, relatively thin components, such as truck frame side rails.

Flexible, programmable fixtures, intelligent sensors, and computerized numerical control: These devices are hardware designed to permit tooling systems to adapt to different body assembly sequences without them having to be disassembled or reconfigured. These systems can recognize the differences in body component shapes and automatically adapt the fixture for proper fit-up, welding, and hole piercing.

All of these new technologies are assisting manufacturers and suppliers to remain competitive or gain a competitive advantage over other less well integrated suppliers. The availability of this advanced tooling allows suppliers to bring new design ideas to the OEMs and actually show them how laser welding can result in more uniform weld performance, significant mass reductions, and a stiffer body structure. In another example, single-step body assembly not only results in better fit between larger subassemblies, but also offers the potential for assembly cost savings and decreasing plant complexity through reductions in end item components. Finally, increasingly sophisticated programmable controllers permit agile manufacturing to become a reality by facilitating rapid reconfigurations of robotic clamping and welding fixtures. Device-level communication standards are necessary so that different controllers can "talk" together, similar to the medium that adjacent workers on the assembly communicate with each other.

As suppliers work their way up the hierarchy of integrated product development, they are typically confronted or rewarded with an increasing amount of design responsibility. As such, the issues that have plagued them will have to be better managed. "Supply-chain management" is a relatively new concept where OEMs assist their tier 1, 2 and 3 suppliers to better manage their businesses. The concept grew out of a concern that without a viable supply base, or "chain," the most profitable, innovative, or well-managed OEM could neither survive nor grow. The process begins with an in-depth, plant-by-plant evaluation of capabilities, problems, and opportunities. Common management information systems, an essential element, are needed to provide not only current production scheduling but also to support broadcasting of components for "just-in-time" assembly. The common database also supports new product development, giving the supplier the opportunity to not just provide parts as designed and fully developed new product concepts but also future product design data. Suppliers who can deliver entire vehicles or large portions thereof are known as "system integrators."

With the opportunity for success comes the risk of failure, which could propagate both up to the OEM and down the supply chain to the end user. The risks, which have always been inherent in new product development, can at least now be shared throughout the supply chain. OEMs have begun to adopt suppliers into their respective "families," for which they receive many benefits other than continued contracts.

8.3 AEROSPACE

All of the manufacturing concerns mentioned earlier are also prevalent in the aerospace industry. In addition, the complexities and quality requirements for aircraft and space vehicles demand even more attention, due to the increased need for occupant safety. Quality improvement, in the context of "zero defects," has been utilized to improve manufacturing processes to the extent that in some systems or subsystems, six sigma acccuracy (six standard deviations from the average) is the goal for manufacturing performance. This can also be interpreted as experiencing less than 3.50 defects per million operations. The standard is applied not only to aircraft and spacecraft, but also to onboard electronics, ground support equipment, test equipment and other infrastructure items (e.g., fixed costs such as the manufacturing plants).

The key operative in this analysis is the application to the design phase. Having manufacturing engineering involved at the earliest possible stage in the new craft design will help to ensure that many of the aforementioned problems will be addressed in the most timely manner. Computer-aided design and engineering (CAD/CAE) software is available to address manufacturing issues very early in the design phase. At the same time, interface programs can translate engineering data into manufacturing commands, thereby eliminating a major source of defects, i.e., misinterpretation of, or mistakes on, the drawings. The interface program will automatically challenge any dimension that is out of tolerance or incompatible with the machine tool to be utilized. Likewise, material selection can also have an adverse effect on manufacturability, if the machinability of a new aluminum alloy is beyond the capabilities of the lathe or mill specified for production. As pointed out in various sources on designing for assembly, up-front design cost represents approximately 5–20% of the total product cost, yet it affects 70–80% of recurring costs, if all of the redesign efforts are accounted for.

Hardware dimensional control complexities have resulted in the creation of an infrastructure and organization not only within the airframe manufacturer, but also including suppliers. This is due to the severe penalties and repercussions for failure. The catastrophe of an airplane crash constitutes the highest level of desirable prevention. Injury to occupants and airline employees would be the next level, but not any less a priority than the first. Primary, secondary, and even tertiary systems protect vital functions including propulsion, flight control, navigation, cabin environment, fire safety, and passenger and crew safety. Multisystem operational efficiency and effectiveness would be at the third level of priority and includes fuel consumption, operational envelope (altitude range payload), load factor, takeoff, and landing distance. Within each of the parameters listed, manufacturing capabilities have to be developed that will guarantee that the above hierarchy is sustained. While failure cannot be eliminated, at least it can be reduced to a level that cannot be consistently measured.

Within the zero defects philosophy, different degrees of fault identification and tolerances must be established. This will allow design trade-offs to be made so that a minimum fault tolerance can be maintained without adversely affecting operations. For example, if a manufacturer of gas turbine jet engines wants to increase fuel efficiency, and therefore range, without adversely affecting short- and long-term reliability, certain trade-offs have to be made. The particular manufacturing environment desired would accept a limited amount of foreign object ingestion without causing a catastrophic turbine blade failure. The limited ingestion scenario would be, primarily, a design issue first, but the manufacturing community would have to be able to achieve the design objective.

Within the airframe manufacturing cycle, every fabrication, inspection, and assembly station has supplier-customer interfaces, both of which have to sign off on the process accuracy and functional performance. There are other systems helping this process, among them relational databases, which contain all product information, performance standards, process capabilities and material specifications. Updated continuously and instantly, these databases allow for continuous process improvement. Outside suppliers are also part of this manufacturing quality system. Industry quality standards are utilized so that common terminology and testing methods conform to worldwide requirements, resulting in completely integrated manufacturing.

While Simultaneous Engineering is frequently associated with overall improvement in product development processes, designing for manufacturing and assembly is directed more specifically toward reducing subsystem elements that do not contribute directly to product functionality. Subassembly components that could be combined to eliminate assembly operations and labor costs are the target. Designing for manufacturing and assembly also includes eliminating left- and right-handed components when interchangeable parts would suffice. Aiding the actual assembly of components includes providing inset or relieved areas on machined parts so that mating parts will naturally "nest" together, instead of reorienting first. High-speed machining of an assembly, out of a solid block of material, may eliminate a costly and potentially unreliable subassembly of thin-gauge sheet skin, extruded reinforcements, and rivets. One rule of thumb, currently practiced, is that if two mating parts do not need to be separated for servicing or repair, do not rotate against each other, and do not contain any other components, they are probably good candidates to be combined in a sub-assembly.

The challenge in any manufacturing environment is to continually question the "status quo" and constantly look for opportunities to reduce cost, without compromising any other product function or attribute.

8.4   HEAVY-DUTY OFF-ROAD VEHICLES

These industrial segments are characterized by multiple-variant, low-volume, high-quality products, for which meeting individual customer requirements, in some cases, results in a high-level of manufacturing complexity and cost. Flexibility has to be the key operative for a low-volume custom manufacturer. Since this type of operation can be classified as a large "job shop," maintaining and improving workers' skills and technical requirements should be considered a top priority. Fabrication and assembly machinery should be specified to handle virtually any size job. Castings and forgings are typically part of these products' component parts, and usually require a high degree of finish work on large lathes and milling machines. Automation in the form of numerical controls is the normal level of sophistication.

Suppliers play an important role in these classes of products. Sophisticated quality control systems are tied into the assemblers' requirements almost exclusively. Military standards, specifications, and tolerances are utilized when no industry standards, such as Society of Automotive Engineers (SAE), are available or applicable.

8.5   ELECTRONICS

Electronics manufacturing generally involves highly automated processes designed to minimize assembly time, manual sorting of components, and uniform high quality. Many integrated circuits incorporate self-testing capabilities for interim as well as final functional confirmation. The basic fabrication procedures were briefly described in Chapter 6. Key design parameters for optimizing manufacturability are listed below:

Manufacturing

Insertion times and costs
Fault rates

Utilization of existing equipment and lines

Quality control and inspection

Cost

Printed circuit board (PCB) layout trade-offs and package style

Separation between ICs

Number of layers and ground planes

Total PCB area, track width, and track area

Ground and signal bounce

Optimization of all of the above parameters requires a large-capacity computer in order to process the millions of iterative calculations necessary to answer all of the questions posed by the resulting matrix. However, there remain other design parameters affecting the ultimate success of any electronically integrated circuit. Input and output characteristics of the upstream and downstream functions will determine the circuit board layout and design as much as will the factors described above. A black-box approach should first be utilized to analyze system requirements. Then each IC can be assigned the functions it has to fulfill. Only after this step can individual IC requirements be planned in light of the above parameters.

8.6   PACKAGED GOODS

The main issue is to prevent the packaging process from altering the product or compromising safety, quality, or performance. Environmental sealing is a primary objective. Some food products need to be protected from exposure from temperature extremes, ultraviolet light, excess humidity, and other conditions. Shelf appearance and product safety could be compromised by degraded coloring, deformed shape, or offensive odors. Recycling of packaging materials has become a major issue for municipalities running short of landfill space. Finally, bar code accessibility is an essential requirement for the package, not only for store inventory control, but also for data used in market research studies, potential recalls, and so on.

Packaging machinery is designed to automatically cut, stamp, fold, form, trim, glue, fill with the product, and seal the package. Frequently, the package becomes part of the product as sales promotional messages, instructions, warnings, and other information are contained on the package exterior. Packaging tooling consists of machines that cut flat cardboard or use precut blanks, preprinted with graphics and text.

Other equipment stretch or vacuum formed, thin, clear plastic film into domes or other shapes to provide see-through capabilities. Some plastics can be shrink-wrapped to provide a relatively tight fit around a product, protecting it from being damaged by its moving inside the package. Frequently, cardboard and plastic are combined to sandwich the product between them, providing a tight weatherproof seal and a place to advise customers on usage, disposal, and so on.

8.7   AJAX MANUFACTURING COMPANY

Much of the manufacturing engineering department's job was to remove as much cost from building the M-2000 as possible, without sacrificing any parametric performance. Nearly 67% of the final assembly cost was manual, with purchased parts only 10% of the total parts cost.

The moving and fixed coils were wound on antiquated machines that were designed, built and maintained by Ajax employees over many years. Several offshore suppliers possessed new coil-winding equipment, which would have resulted in improved quality but potentially unfavorable delivery times for Ajax. All of the domestic coil winders, in response to imports from foreign coil manufacturers, had been either forced out of business or had turned to a less competitive and more profitable product.

Purchasing found another alternative in the form of a used, domestically built coil-winding machine, that was available at a third the price of a new machine. The business case analysis that was performed by the engineering staff could not justify the cost of the machine, based strictly on volume, unless the company chose to also get into the custom coil-winding business as a sideline. On the other hand, without a reliable supply of coils, production of the M-2000 was in jeopardy. In the end, the decision was made to purchase the used machine, with the potential to market custom coil-winding, on short- or medium-volume quantities, to local or regional customers, competing on delivery time.

Another crisis occurred when the tolerances for the tracer tube, which contained the fixed and moving coils, could not be met. The inside diameter had to be maintained within ±0.0005 in. in order for the tracer output to remain linear. The lathe utilized for turning the tube could not hold this tolerance consistently, resulting in a high scrap rate and driving up the manufacturing cost. The prototype units were fabricated by toolmakers utilizing an older but well-maintained instrument lathe. For volume production, a faster, numerically controlled lathe was to be used. Ajax Manufacturing engineers pondered their options: rebuild the existing lathe, buy a new machine, or farm the job out to a subcontractor. They opted to send the job outside until the lathe could be rebuilt.

9

PROMOTION AND DISTRIBUTION

9.1 INTRODUCTION: COMPLYING WITH ALL PROMOTIONAL AND DISTRIBUTION OBJECTIVES

Promotion is all of the marketing activities involved with getting customers to purchase the product. It consists of several distinct facets: advertising, designed to generate interest in and demand for the product; graphic arts, to develop print and packaging themes; and multimedia relations, to coordinate the release of materials at the time of product introduction. Several design-related considerations should have already taken place, on a new product program, prior to this phase of the process. Market research would have identified key customer groups, performed market trend and demand analyses on similar or current product versions, and analyzed the results of focus group evaluations of beta tests on prototype or early production units. Another item that must be determined is the extent to which the new product affects or is affected by the brand strategy.

9.1.1 Brand Management

Brand management is a relatively new concept for some industries. While it has been an essential element of the marketing strategies for consumers goods for many years, it is now being explored for other products as diverse as automobiles, insurance, and financial services. The theoretical aspects of brand management should address all of the factors that could or do have either a positive or negative impact on the brand identification and/or image. The brand generally consists of a company or product name, a symbol, an advertising slogan or promotion, animation, graphic artwork, logo, and other elements. The brand, along with all identification, is registered as a trademark or copyrighted material. Brand management also extends into planning how the brand identification can be consistently utilized for marketing advantages, to maintain or improve share, profitability, and so on.[1] It also involves how the brand interacts with other brands, both inside the company and with competitors.

When merger and acquisition activity cause brand proliferation inside the company, the potential interaction can be detrimental to all of the brands involved. In recent times, it is not uncommon to see previously competing brands still competing, but under the same corporate umbrella. Brands can also have life-cycle properties that also need to be considered, although some consumer product brands are in excess of 100 years old. Frequently, competing brands within the same corporate entity are not be allowed to coexist for an extended time. An analysis should be performed to decide the relative strength and fit of each competitor, resulting in elimination of the weakest members of the group.

When new products are being planned, a brand strategy must be developed if branding is a strong element in the marketing plan. The strategy should entail, or at least not be in conflict with, all of the corporate themes that were developed in the strategic planning phase. If the brand is envisioned as a virtual extension or substitution of the corporate "presence," it should be noted that tying the two together closely can either be good or bad depending on how well both of them fare in the market. If the brand image of a product has been one of highest quality, price, and performance, a new product, responding to changing market demands and having a lower price and performance, may be difficult to position or align to the same brand as the previous product. A new brand may have to be created for the new product.

9.1.2   New Brand Design Issues

Assuming a new brand is needed, the following should be considered during the design phase:

Product's functional characteristics

Intended utilization

Target market

---

[1] Aaron Robinson, "GM brands born again," *Automotive News*, January 4, 1999, p. 27.

Research results from clinics on "intrinsic" product values experienced by respondents

Potential advertising themes from idea generation brainstorming sessions

Export market considerations (name recognition without any negative connotations)

9.1.3   Promotional Plan Development

A successful promotional plan will need to address not only the brand hierarchy inside the company and the product's market position, attributes and performance characteristics, assumed or acknowledged warranties, among other features, but also the relationship with the media, in all of its forms. Coordination between market introduction, print advertising, electronic versions, and advanced dealer or distributor involvement are all issues that need to be well thought out ahead of time. Significant advertising, media plan, and other promotional parameters are listed below:

Advertising agency selection

Cost (creative copy generation, media transformation)

Timing (frequency, market-segment exposure)

Media channel selection (print, radio, television, Internet, billboards, etc.)

Content approval (corporate and brand legal staffs)

Trademark registration and copyright verification

Assuming the new product program had been designed to meet the above criteria, design and delivery of promotional materials should be easily accomplished.

9.2   DISTRIBUTION

Distribution involves all aspects of moving the product from the manufacturer's plant to the end user, by way of the transportation system, warehouse or storage facilities, inventory management, and accounting systems. Direct sales expenses should also be taken in account since they can affect or influence the distribution outcome.

9.2.1 Transportation Issues

Transportation involves physical movement of the product to the various market venues. More than 50% of all manufactured products move to either their intermediate or final destination via heavy-duty over-the-road articulated tractor trailer trucks. Many are finished or work in progress goods from suppliers going to final assembly or distribution centers. Most heavy, semi-finished, or finished goods and raw materials are transported by rail, barge, ocean, or inland waterway. Air shipment and trucks are widely utilized for time-dependent or critical commodities needed for emergencies or other purposes. Costs are proportional to time in transit, or distance traveled, and weight, since weight directly affects fuel usage. Costs are charged whether the goods are moving or not; therefore, shippers are constantly striving to improve on time performance. Loading and unloading costs vary widely depending on the degree of automation in the given industry. Tracking of some trucks, barges, and ships utilizes ground position satellite–based navigation systems, while most rail movements are centrally controlled by remote, computerized dispatchers. Mergers and acquisitions can have a dramatic impact on these operations, as will be shown later.

9.2.2 Warehousing and Inventory Management

After delivery to the distribution point, warehousing and inventory management systems are utilized to track the items until they get shipped out to the wholesale and retail outlets. Some products, such as automobiles, are shipped directly to retail dealers, for delivery to customers, after the necessary preparation and cleaning. Usually, products ordered by the retail establishments are then shipped from the distribution warehouses. With the advent of advanced inventory management software and bar coded packaging, goods in some industries—notably some food products, household supplies, sundries and other high-turnover products—are shipped directly to large retail store chains from the factory, thus eliminating the warehouse/wholesale steps and the costs involved.

Certain aspects of distribution networks need to be analyzed in case product design dependencies are uncovered. The nature of the product will determine which distribution channels are optimum. Size, mass, volume, cost, and delivery time to the end user are all important considerations. The distribution system wherein a manufacturer sells to a distributor/ wholesaler, who sells to a retailing dealer, who in turn sells to a customer, has evolved over many years to an assumed-to-be optimum situation.

When some businesses were small, direct selling was undoubtedly more common but very inefficient. Later on, higher volumes of mass-produced, nearly identical products dictated a different system. Market demands for product differentiation resulted in a proliferation of product line extensions, functional variations, and export versions with foreign language instructions and markings. Finally, in some mass production industries, manufacturers are evaluating the concept of virtually unique products, modified to each customer's wants, built by a special assembly team, and shipped—all within three working days. Assuming that the design, engineering, and manufacturing problems are solved (or at least, solutions for them are predicted ahead of the order receipt), the current distribution systems would at least double or triple the three-day target time interval. While it is conceivable that the typical bottlenecks in the system can be eliminated through extraordinary means (i.e., expediting, express air freight shipping, and so on), the added cost would have to be amortized over a much larger base of the products, not requiring the added attention. Competitive pressure will require eventual reductions in other costs, in order to remain solvent. This is another reason for minimizing product costs at the design stage, and at every phase of the new product development process.

9.2.3 New Distribution Requirements

The main reason for the above concept evaluation situation is the development of new promotional, transactional, and distribution channels, mainly through the Internet. It is possible to have a new private passenger vehicle shipped from a regional storage facility on the same day it was purchased and deliver it to the customer two days later. This has whetted the appetites of both consumers and marketers for further explorations. Future developments may result in three working-day deliveries of new vehicles, customized to the buyer's requirements, through a terminal at the dealership or elsewhere. Obviously, the distribution side of the equation will have to change radically. In the case of motor vehicles, most long-haul deliveries are made on specially designed rail cars, with elaborate tie-down systems of clamps, chains, and other restraints. Special express trains would have to be created for moving high-priority freight, a distinction reserved until now for perishable foods such as fruit, meats, and vegetables. Most loaded vehicle rail cars stand for typically two to three weeks in marshalling yards, either adjacent to the assembly plant or at the rail head, waiting for a train that is going in the same general direction as the vehicles' destination. Based on an analysis of current rail system problems, increased traffic—

due to a generally robust economy, decades of cost reductions through selling off of then excess capacity, and mega-merger activity—has left the rail industry barely able to keep ahead of demands for increased service from regular commodity customers, without considering the special requirements of future vehicle shipments.[2,3] Experts predict that it will cost billions and take years to enable the rail system to eliminate bottlenecks and restore and improve customer service for the current freight mix.

9.3 AJAX MANUFACTURING COMPANY

The traditional method for promoting industrial instruments was through manufacturer's representatives or independent distributors, providing the market demands warranted it. Trade shows were also frequently utilized to introduce a new product, allowing potential buyers to observe live or videotaped demonstrations, or try out early production devices themselves. Then, after such time passed as was necessary for production units to be available, the sales reps would begin making the rounds, talking to their regular customers, discussing new sales literature, and answering any questions, all the while asking for orders.

Quite often, another time lag from several weeks to many months elapsed until customer orders started coming in. Many manufacturers gambled on the hunch that, once regular production began, the costs associated with building an inventory of "standard" units could be partially offset by being able to ship filled orders very fast, offsetting any that might be canceled due to customers not willing to wait for production delays. Ajax senior management realized that something different was necessary for the M-2000 to make any inroads on the established brands and models. While the product was still in the concept stage, the marketing vice president held a meeting to brainstorm ideas. Among the ideas generated was to "assign" early production units to established customers and let them operate the devices for up to a year, in a free trial without charge. The sales reps would carefully monitor the results to make sure any problems or questions were being adequately addressed. At the end of the year, or before the trial period was up, a regular production unit would be brought in, and the trial instrument would be sent back to the factory for evaluation.

---

[2]Bill Stephens and Craig Sanders, "Cleveland: Center of Controversy," *Trains*, July 1998, pp. 24–27.

[3]Don Phillips, "Railroad Customers Are Just So Lucky," *Trains*, February 1999, pp. 14–15.

The advantage was in winning the customer's attention and getting technicians to operate the device on a regular basis, to build confidence in its accuracy, speed, and other features. However, the potential disadvantages included losing technicians' confidence if there was a breakdown or other problem that took the unit out of service. The proposal included having backup instruments available at a moment's notice. This meant that reps needed to be available, virtually around the clock, since many automotive customers operated three full shifts per 24 hours. While the reps were not keen on this part of the plan, they went along with it since it was seen, by the more progressive of them, as an opportunity to promote other product lines. If the M-2000 was seen to be in nearly constant use, then other instruments would probably get more usage also, since it was a sign that more measurements were required across the board.

The proposal was not considered for adoption immediately, but was tabled until the bugs were worked out of the new product's design. Later, as the new instrument development neared completion, the proposal was reconsidered and approved.

Internet sales represented another new opportunity for Ajax. Animated demonstrations and customer testimonials were developed, along with creative new sales "literature," which took the potential buyer through a virtual reality simulation of being on the shop floor, measuring machined parts, and printing out the results. Special computerized animated "operators" or inside sales personnel received inquiries and orders over the computer. Any technical questions were referred to the engineering department, where 24-hour coverage had been established several months prior to the introduction of the M-2000. Order processing, credit checking, and other operations were all handled without human intervention, unless required to correct a problem or by customer preference.

10

AFTER-SALES SUPPORT, MAINTENANCE, SERVICE, AND REPAIR

10.1 INTRODUCTION: CERTIFYING THAT ALL MAINTENANCE, SERVICE, AND REPAIR REQUIREMENTS CAN BE MET

After-sales support generally comes from factory-authorized and independent businesses, established to maintain both the product and contact with the customer. As costs of product ownership have become more significant, factory warranties have also assumed their prominence. Factory-authorized repair stations have the benefit of corporate support, although most are independent businesses. Franchised dealers also provide service parts and repair service.

Nonoriginal equipment parts are sometimes available through independent suppliers. This practice most often occurs when manufacturers contract with outside firms for black-box subsystems and components, which could include extended warranty coverage, maintenance contracts, and out-of-warranty repairs. Maintenance contracts between manufacturers, distributors, and dealers provide consumers with an alternative or a supplement to factory warranties. These devices are basically insurance policies to protect the consumer against having to pay for a potential product defect or failure. Warranties are not insurance against all failures and can have limited market value if the product experiences chronic field problems requiring repairs.

121

10.2   PRODUCT SERVICEABILITY AND PARTS

Product serviceability guidelines are utilized by engineers and designers to provide space, techniques, and tools to permit service operations to be carried out as efficiently and cost-effectively as possible. These guidelines represent the culmination of many hours of analysis, component removal, and reinstallation trials, time studies as well as other efforts to minimize the problems encountered by service technicians in the field.

Service parts operations are established in order to insure that adequate replacement parts are available at a reasonable cost and in a timely manner, through distribution centers or warehouses, located in strategic areas of each market. Some industries have a 10-year replacement products policy, although it may not have any basis, legal or otherwise, except for customer satisfaction and convenience.

Automotive serviceability guidelines include providing accessibility to the most frequently replaced or potentially warranty-sensitive components or subsystems. Optimum accessibility means not having to remove any other component prior to removing the one required. Special tools and procedures may be required but should be minimized, except for franchised dealers. Accessibility also includes providing sufficient clearance to utilize hand tools.

The following topics provide after-sale information on four of the five industries studied. Packaging materials are either disposed of or recycled.

10.3   AUTOMOTIVE REPAIRABILITY

Automotive repairs are classified as either mechanical and electrical service and/or component replacement or body repair. Mechanical and electrical repairs are generally handled by new car dealer service departments until the factory warranty expires. Then, independent garages continue servicing the vehicle until it becomes economically nonrepairable and it is scrapped. Body parts are generally performed at independent repair shops, and those involved in direct repair programs are administered by the insurance industry.

10.3.1   Automotive Body Repairability

Body repair techniques for collision damage are a separate situation from the factory warranty one. Most passenger cars have unitized welded structures that are designed to crush for optimum occupant safety. The built-in safety systems are primarily designed to activate at closing or barrier speeds of 20 mph or higher. However, the need to protect unbelted occupants has led to the establishment of relatively low thresholds of 8–11 mph for deployment of air bags, in some models. The need to trigger the air bag system typically involves creating a crash pulse in the front structure as it collapses from a frontal collision. If the energy level is sufficiently high to require the air bag deployment, to supplement the belt system or protect the unbelted occupant, a sensor detects that the deceleration is above the threshold and closes a switch. A second switch also has to determine whether the impact is severe enough for an air bag to be of assistance. Finally, a deployment occurs.

Unfortunately, the ability to restore the vehicle to its preaccident condition is not as predictable as the crash outcome. Once the unitized structure has been deformed beyond its elastic limit, the material typically utilized (high-strength, low-alloy steel) cannot be returned to its former state without a significant strength loss, approximately 35–40%. Therefore, the only repair alternative is to remove the deformed portion and replace it with an undamaged piece. Gas metal-arc welding techniques have been developed that replicate the yield strength and fatigue resistance level of the factory spot welds. Corrosion resistance has been shown to be equivalent to that applied at the plant, although the material is different. Subsequent accelerated durability and impact tests have confirmed the restored strength level of the unitized body structure, to within ±10% of the original construction. Restoration of corrosion protection remains problematic since the factory dipping and coating procedures can only be approximated in the field.

While it is statistically infeasible to account for all possible impact direction and speed combinations, the global motor insurance–sponsored repair research community has devised a series of low-speed (9.4 mph or 15 kph) offset barrier tests for driver side–front and opposite rear-corner crash tests, as a measure of overall damageability. Insurance rating mechanisms have been developed in some markets in Asia, Europe, and South America based on this test series along with other pertinent information of new materials and manufacturing processes that could influence repair complexity and cost. Comprehensive claim experience is also as important as collision, since theft is the single largest component. With the proliferation of factory-installed immobilizers, theft is somewhat easier to prevent than the haphazard methods of the past.

10.3.2   Vehicle Design Recommendations for Repairability

Design recommendations on optimizing damageability and repairability should consider the following the basic premise concerning low-speed collision and theft damage:

Collision damage should be contained to the outboard areas of primary structural components (e.g., rails, pillars, etc.).

Repairs should be facilitated through ease of access or for removal and reinstallation of components.

Partial replacement procedures should be developed and available at the time of market introduction.

Replacement parts configuration should facilitate ease of installation.

Theft deterrence should include both content and drive-away counter-measures.[1,2]

10.4 AEROSPACE SERVICE AND REPAIR ISSUES

The airline and aerospace industries have developed a highly sophisticated system for insuring that airliners get regularly scheduled maintenance on all systems, including daily and other periodic inspections, jet engine over-hauls, operational checks on navigation, communication and monitoring instrumentation, spare parts supplies, and so on. Service and repair information is typically supplied, along with the components and systems, by the manufacturer.

Maintenance engineering is a formidable and credible group when exposed to a new airplane design, particularly when their input was not previously sought. After-the-fact consideration of maintenance issues has resulted in significant expenditures in redesign time, when tool clearances and logical recommendations for spare-part configurations were either forgotten or ignored. Most aerospace firms have elevated maintainability to a high enough level to positively influence advanced designs at the earliest possible times.

10.5 HEAVY-DUTY VEHICLE SERVICE AND REPAIR ISSUES

Among commercial vehicle manufacturers, maintainability tends to command a prominent spot in new vehicle design activities, since the equip-

---

[1]*Optimizing Damageability, Repairability, Serviceability and Theft Deterrence*, Recommended Practice J1555, (Society of Automotive Engineers, Warrendale, Penn., 1985, rev. 1993), p. 1.

[2]*Vehicle Design Features for Optimum Low Speed Impact Performance* (Research Council for Automotive Repairs, Stockholm, Sweden, 1995), pp. 1–3.

ment is typically amortized over a 10-year life, requiring that it be operable for at least as long as the depreciation schedule, and usually longer. Dealer service organizations are highly developed and generally successful. Parts distribution networks have been in place for a long time.

With the advent of imported vehicles and systems, the previous network is being strained to maintain a proliferation of new models, many of which are relatively low in volume. While maintainability has not attained the status of the aerospace industry, most manufacturers have their own staff working on these issues. One suggestion, which has been offered repeatedly, is to assign new, young engineers to the service and repair staff, for a minimum of five years, in order for them to experience, firsthand, maintenance problems that could have been avoided with a product better designed for service and repair.

10.6   ELECTRONICS PRODUCT SERVICE
AND REPAIR ISSUES

Most PC-based products have a somewhat modular design, so that separate RAM, hard drive, and other components can be replaced when defective. If integrated circuit boards fail, it is usually more cost-effective to replace them than attempt repair.

10.7   AJAX MANUFACTURING COMPANY

One of the reasons Ajax management decided to proceed with the M-2000 was the positive experience Ajax had with repairing other manufacturer's equipment, often with a faster turnaround time and fewer total problems than customers could achieve by involving the other manufacturer's service people. Since they had developed their repair business into an entity worth 25% of Ajax's total annual revenue over the past 15 years, senior management had eagerly sought to exploit their advantage by purchasing older competitive units, refurbishing them, and providing one-day replacement service—all for new as well as older customers.

As the M-2000 project neared completion, Ajax senior management contemplated how the new device would impact the service and repair business that was firmly entrenched and working well. They worried about disrupting the established routine to the detriment of existing service arrangements. Though meetings were held with the service staff from the early stages of the new product development process all the way to the present, there were still some misgivings on the part of technicians. For example, the methods of construction, assembly, disassembly, and calibration were entirely different from those of the old pilot mechanism. Components for the new device cost more to manufacture and took longer to install than those for the current products.

Electronically, the machines had similarities, but mechanically, they were completely different. The main concern was that the current service staff could not provide the same or a higher level of service as what the new customers would come to expect, based on the experimental quasi-beta test introductory period.

Senior management rightly understood their concern and asked the technicians for their ideas for alleviating the potential problem. The brainstorming session produced several interesting concepts including a separate service department for the M-2000 and a service contract management system that would permit on-site repair of damaged or out-of-calibration units.

The management team analyzed the potential favorable and unfavorable points of each suggestion, making their decision after several months of negotiations with financial advisors and technicians. Finally, the announcement was made: There would be a separate service department for the M-2000, but only for a year, allowing for a gradual phase-in for the rest of the service department. On-site repair services were also to be offered on a limited basis for selected customers, and funded from a set-aside from M-2000 sales. All technicians were eventually to be involved in the "road trips," as they were dubbed. Scheduling was somewhat hectic at first, since there were still some operational problems with early M-2000 units, until all were upgraded with newer, more stable electronic modules. After one year, full service integration was achieved after additional technicians were hired and trained through the production process.

SECTION II
CASE STUDY EXAMPLES

Each chapter in this section provides a generic case study example, one for each of the five industries under examination. The case studies are generic to protect the confidentiality of proprietary information and to increase the time value of the information. Each industry highlighted has a new product example based on factual information, either directly experienced by or communicated to the author, supplemented with fictitious scenarios to more graphically portray, as close as possible, the most realistic, but also idealistic situation.

Each case study follows the same format, starting with the sequential phases of the 10-step new product development process, and exploring how Simultaneous Engineering considerations provided for efficient and effective decision making at each step along the way.

11

NATIONAL MOTORS CORPORATION

11.1 INTRODUCTION: AN AUTOMOBILE MANUFACTURER DECIDES TO UPGRADE ITS PRODUCT LINE

The chairman of the board of directors and chief executive officer of National Motors Corp. just received the latest monthly sales report from the vice president of marketing. The report listed the last 30 days' selling results for the firm's current models along with projections for the remainder of the sales year. It showed not only a continuation of a decreased selling rate, first recorded several months earlier, but also a steep decrease or deceleration in the rate.

A report from the marketing committee meeting was attached to the sales report. An analysis of the sales decline listed each model in the current lineup, along with competitive vehicles, both imported and domestically produced. National's leading model, the Centaur, a subcompact four-door sedan, had been completely redesigned only two years before. Since last year's introduction, in which sales had been encouraging, the sales results for this year had been flat earlier and now were seriously decreasing. While the report contained mostly negative news, the analysis showed that the competition was generally faring no better. The top three subcompact-sized models all had decreased sales ranging from 5–22% relative to the prior month, and an average of 14.7% compared with the prior year. The model with the highest market share, the XME 350, an all new car introduced last year, had the worst monthly performance of its short product life, a fact that had already caused a virtual panic in the manufacturer's headquarters, in another country. The other competitors were facing either slight decreases or flat sales relative to last year.

The sales analysis continued with the major reason for the trend: increasing demand for other types of vehicles, in particular, light-duty trucks of all kinds, sport utility derivatives, and even large passenger cars. The demand was fueled, literally, by the price of gasoline, at the time cheaper than bottled water, and headed lower later in the year, based on an oversupply and a lack of economic growth in other markets, particularly in the third world. Overexpansion of oil production, in anticipation of increased worldwide demand had also contributed to this condition.

Fuel consumption conservation efforts in the Western countries was beginning to have a positive effect, with the building of significant interest in mass transit systems and the exploration of alternative energy sources underway. These efforts were more the result of alarm over increased fossil fuel emissions and reduced air quality than by fuel consumption concerns. Ozone alerts had plagued virtually every major world city during every summer season for the past five years, and the health threats were getting worse each year, based on an increase in reported cases of lung and breathing problems. Acid rain, from sulfur dioxide emissions, was also increasing. Finally, interindustry efforts to develop new fuel-efficient designs and innovative manufacturing methods were underway in several countries, but were years away from production.

The Centaur had been redesigned partially in response to the concerns recounted above, i.e., a small car with reasonable interior space and utility, and a power plant engineered to produce extremely low emissions and high fuel economy. Initially, the public interest and acceptance of the vehicle had been overwhelming. Advance orders initially poured in from dealers, which usually had trouble placating buyers not willing to wait a year or more for a new car. Now, since the anticipated increase in public demand for lower emissions had moved from the front pages of newspapers to the back page, interest in the Centaur had faded along with the news. Many of the advanced orders had been canceled, resulting in an inventory backlog of over 80 selling days of supply.

The CEO, as chairman of the strategic planning committee, called an emergency meeting for two days later at 1:00 P.M., in the large conference room on the top floor of the headquarters building. Normally, invitees included the vice presidents (or their designates), of marketing and sales, controller's office, design, engineering, purchasing, public relations, legal, manufacturing, research, and service. However, this meeting was mandatory, which allowed no alternates or designated substitutes. Two vice presidents had been on vacation in the Caribbean and California, and another on a business trip in Europe; all three were called home.

The atmosphere in the dark wood-paneled room—which lacked windows in order to minimize distractions, and was appointed with rich and heavy wall hangings, somber portraits of past officers, and massive, if immovable, furnishings—reflected the attitudes of the occupants. All of the invitees knew what to expect, since this kind of meeting was not utilized for celebrations. All of them had been anticipating it for several weeks and had been making preparations. This usually involved some second-guessing on the part of the staffs. They were concerned not only with the latest sales report but also with the general lack of new offerings by competitors, the anemic economic reports, and the usual political infighting that seemed to have the markets foundering in uncertainty for the last three months, as if the analysts and investors were waiting for some major new crisis to descend on the scene. The international situation was also quiet, with only a few brushfire wars occurring in several remote areas, and all of the major petroleum-supplying countries were temporarily stable.

All participants were seated by 1:00 P.M., as the CEO entered the room and took the chair at the head of the large oval table. He greeted everyone warmly and looked over his notes for a few moments. "Gentlemen," he said, "I would be most appreciative if you could offer us a solution to the problem with the Centaur. As you probably know by now, the continued sales slump may cause the plant to be closed for up to three weeks per month for the near future. The rest of the product line isn't doing much better. Orion, [a midsized luxury sedan], has a 49-day supply, while Pollux [a sporty two-door hatchback], has an 87-day inventory."

The marketing vice president spoke first: "Chief, I think we have to face the fact that the marketplace has fundamentally changed. For a niche manufacturer, without any truck capacity, or resources to acquire one even if it was available, we are going to experience these kinds of problems, at least in the short term."

The controller interrupted the marketing VP. "Excuse me, but I believe the problem is more than just short-term. As you said, the market has changed, I think, permanently, toward bigger vehicles, just as the market research report stated. There is no indication that crude oil prices will increase anytime soon, and if they do, it won't be by much, since there's more oil available now than there was in the last 40 years, and its cheap."

The purchasing VP then said, "I know an Asian manufacturer who will sell us 100,000 small pickups, Centaur-badged, two- and four-wheel-drive models, for $5300. and $5800., respectively, F.O.B. any West Coast port."

"So what!," retorted the Marketing VP, "none of that small stuff is selling anyway, so why should we invest almost $6 billion in another dead-end product line!"

"All right, gentlemen," said the CEO, "let's at least agree that we need a new product plan as soon as possible. I want a new design that will demonstrate our superior engineering skills into an entirely breakthrough concept, but one that fits the needs identified in our research clinics. I expect everyone to put their best people on this—I want to see some results quickly. We will reconvene here in three weeks, and I will expect to see sketches, preliminary layouts, and printouts on projected performance figures, mass, timing, and costs for major systems. Also, I will be looking for any backup data you have on the competition. I have an appointment with the board of directors in six weeks, and I want to request funding for a new product program at that time. I know you will not let me down; we've gotten through worse times before, and this is no different. Any questions? No? OK, good day, gentlemen."

11.2 NEW PRODUCT COMMITTEE

The next few days were hectic as the new product committee got organized. The first meeting took place at 7:00 A.M. the next day. The same group as the previous day, but without the CEO, met and decided on various assignments and work tasks. Four new product teams were established, one for each of the concepts proposed by the design office. These teams, designated 1, 2, 3, and 4, were staffed by personnel, selected by the committee, based on the individual's talents and interest in the particular concept being considered. Supplier participation was also included on the same basis, with a strong weighting toward an ability to develop the concept into a production vehicle. Each team had functional as well as staff representation. The functional representatives included some from the prominent engineering departments, based on the lead time required to develop new technology: power train, fuel systems, body design, steering, suspension and brakes, materials, testing laboratories, quality assurance, cost planning, and so on. In addition, other staff personnel were assigned as liaison to other departments, who had interests in each team project but limited direct involvement except for specific or future project phases. The new product development process called for these groups to be included in every meeting involving their particular area of expertise and/or responsibility. These included product planning, government affairs, dealer relations, and interior systems design, among others. Participation in new product committee activities was also considered a top priority and a potential career enhancement opportunity. However, the failure of a new product could potentially have the opposite effect, depending on the amount of lost revenue and profit.

Team leaders were chosen by senior management on the basis of previous project management experience, skills, qualifications, and abilities, generally acquired through the pursuit of an engineering career but not strictly limited to that discipline. Generally, crossover career development was encouraged by the company, with frequent assignments in manufacturing, purchasing, sales, and service, among other departments. These assignments typically required a minimum of five years' experience within the firm. However, recently hired engineers were often selected if special skills were desired. Since project managers often had to negotiate with line chief engineers in order to obtain additional resources and talent, they often came from the ranks of senior engineering managers with 10–20 years of project experience. Some project leaders had as little as five years with the company, yet demonstrated sufficient talent to be promoted over older and more experienced engineers.

11.3   NEW PRODUCT CONCEPTS

The new product committee met first to discuss the four new concepts presented by the design office. Team 1 received and accepted concept C-147, a small wagon derived from the Centaur platform. Team 2 got D-561, a sport utility/wagon hybrid off the Orion platform. Team 3 was given a midsized pickup truck, T-205, all new, without any previous frame platform experience in the company. Team 4 received a unique assignment, an all new design for an adaptable platform, one with some traits of the other platforms but with increased utility. The UAV-100 had features usually found on sport utilities, minivans, and small pickups but combined into one package. The adaptability came from easily configurable modules that could be modified for specific uses by the owner. Detailed mass and cost distributions of the four concepts are shown at the end of the chapter, along with benchmarked competitive vehicles.

The four concepts were prepared by the design staff as a result of regularly scheduled meetings of the new product committee. When not specifically engaged in "active" programs, as the four new designs were considered, the committee met to discuss ongoing developments in market research, innovations from the research department, new proposed government regulations that could potentially affect vehicles, consumer and warranty satisfaction surveys, and competitive actions, among other topics.

The committee was expected to have a number of specific new programs "in process" at all times, in case market demand required a vehicle not covered by existing product lines. This was in addition to annual updates of current platforms, such as next year's Centaur, Orion, and Pollux. The research department had the ongoing responsibility of monitoring developments in other industries that could provide benefits to the firm. Purchasing had an ongoing assignment of encouraging suppliers to develop their own new product concepts and present them for review to the engineering staff, on a quarterly basis.

The results of all this activity culminated in the new product committee meetings. The first meeting after the CEO's announcement was chaired by the marketing VP. He spoke first to the team leaders to find out if all their participants were onboard, informed, and motivated. All four were able to give progress reports on the various details for each platform concept. Teams 1, 2 and 3 leaders predicted that they would have most of the requested data available for the deadline, in less than three weeks. The leader for team 4 hedged on the schedule, due to unfamiliarity with a new exterior panel material, a superplastic composite, which had a potential for both weight and cost savings, advantages for molding, superior toughness, and resiliency for resisting minor collision damage. The supplier for the material was conducting tests for fatigue and aging effects, the results of which would not be available for at least 30 days. The marketing VP told him that a decision to utilize the new material might have to be made prior to the test results in order to maintain the product development schedule, however risky it might appear. The team leader thought for a minute and stated that earlier test results might suffice if the supplier could furnish a correlation between the material variations tested. He said he would pursue this point with the supplier. However, he knew that follow-up testing would have to be performed as soon as possible.

Two weeks had now gone by. Team 1 had perhaps the easiest job since a wagon version of a four-door sedan amounts to extending the roof and changing the decklid into a liftgate. Remaining to be done were interior noise and volume measurements, roof rack design, mass and performance calculations, and modifications to the rear floor, axle, and rear suspension to accept the four-wheel-drive option. A slightly larger displacement, twin overhead cam engine was deemed a minimum to meet adequate performance requirements, as indicated by the marketing research clinics conducted on current-model small utility all-wheel-drive wagons. This market segment was growing in spite of the fact that most drivers utilized the four-wheel-drive for less than 5% of total driving time, considering both off-road and weather-related conditions. Research further indicated that at least 25% of the target market drivers prefer a V-6 engine over an in-line four-cylinder power plant. The main problem was providing enough under-hood space for the larger engine, along with larger accessory drive components, i.e., alternator, power steering pump, and air-conditioning compressor. The added mass would also adversely affect fuel consumption, which could reduce the company's corporate average fuel economy figure, mandated by the EPA and weighted by sales volumes. This meant that slow-selling high-mileage cars could limit the number of low fuel economy truck derived vehicles that could be sold.

Team 2 had a slightly more difficult assignment than team 1: a sport utility variant of the Orion required virtually an all new floor pan to accommodate a larger in-line V-6 engine, four-speed automatic transmission with full-time/part-time drive transfer case, multilink front and rear suspensions with antisway bars, new steering and brake systems, and so forth. The center of gravity is higher for sport utilities, to provide additional ground clearance. This would require a more sophisticated suspension system, not only to control higher body roll but also to counteract other unstabilizing forces that could occur during evasive driving maneuvers, such as quick lane changes to avoid collisions with other vehicles, animals, or road debris.

Team 2 had benchmarked several newly introduced small sport utility models with the above features. Mass, manufacturing cost, and complexity studies were in progress for the underbody mechanical systems mentioned above. Another issue requiring resolution was whether the vehicle frame should be separate or integrated into the body. The major trade-off being decided in the market was whether a sport utility derived from a pickup truck was worth more than the premium charged for a unitized-body sport utility that looked like a truck but felt and operated more like a passenger car. The more significant trade-off, being debated within the industry, was whether or not the added mass (and resultant loss in performance and fuel economy) of a separate frame was worth the incremental decrease in noise, vibration, and harshness (NVH). Most frame vehicles need the added mass for rigidity when hauling maximum payloads. Most small sport utility models have limited payload capabilities, other than passengers and luggage. If more payload or trailer towing capacity is necessary, most customers are directed toward larger sport utilities or pickup trucks, which typically shared major components, such power trains, frame characteristics, and front-end sheet metal.

Team 3 followed the developments of team 2's efforts closely, since team 3's project could probably benefit regardless of how the frame/ unibody debate was resolved. Although most small pickup trucks had separate frames, this was usually the need for sharing design similarities and components with higher gross weight and maximum payload variants. Market research again showed that most customers never utilized the maximum load-carrying capabilities of their vehicles. However, buying patterns indicated a demand for vehicles with larger engines, capable of transporting heavier loads, than was currently available. Unfortunately, National did not have the resources to design or manufacture larger engines, nor were there sufficient quantities of other manufacturers' engines available, since worldwide demand was constrained by increasing supply problems. As a result, both teams 2 and 3 had parallel designs ongoing, one with a full frame, the other with a unitized body, but both utilizing the same 3.0 liter V-6 engine.

Team 4 planned on also sharing the results of the above debate. The removable or modular components were required to have a separate structure underneath in order to maintain rigidity. A separate frame was seen as a necessity for controlling body twist, after roof removal. Major crashworthiness concerns had been identified early in the design phase. Various schemes were being contemplated, including slip joints that bolted together at major structural intersections; extruded seams between exterior panels, to control fit-up; and a reinforced rear floor pan with rail groves for bolting down rear seats, cargo, and so on. The concern was over the customer's ability to perform the various modifications correctly, everytime, without jeopardizing occupant safety. Removing and reinstalling the roof-mounted cargo pod was the least sensitive, with four locking pins being envisioned as sufficient. Removal or repositioning of seat assemblies, in glide tracks similar to aircraft seats, was considered acceptable. New seat suppliers were demonstrating seats with integrated tracks and belts at recent trade shows.

Operation of the vehicle without the rear roof/cargo area cover was judged to be somewhat risky but acceptable. Due to the potential of a rollover, roof crush resistance would need to be verified as part of the safety standards compliance. Also contemplated but considered somewhat risky by the team was an idea for a variable wheelbase, utilizing a set of telescoping rear rail sections with overlapping lower side panels and rear floor pan, movable rear wheelhouse corner modules, and flexible brake hoses and fuel lines. Interlocking, removable intermediate upper side and roof panel sections would be feasible with the rear roof removed, like a pickup truck with an open cargo box, the rear liftgate removed and replaced with a tailgate (or dutch doors).

Power train options for each concept were limited. The standard 2.0 liter single-overhead cam (SOHC) and 2.5-liter double-overhead cam (DOHC) four-cylinder engines were adequate for the Centaur, but would be severely taxed by the weight and payload increases planned for each new concept. A 3.0-liter V-6 was available in limited quantities from another manufacturer, but doubling the quantity would require a new engine block and cylinder head casting facility and assembly line, estimated to cost $4.5 million. Transmissions were less of a problem with both four-speed automatic and five-speed manual gearboxes available from the regular Centaur supplier, with no quantity limitations. Supercharging the V-6 was under analysis to test the strength of the crankshaft to accommodate the added stresses.

Since Centaur was due to receive the V-6 at the next model change, team 1 had another advantage over the others, inasmuch as the layouts and engine design room mockups were in progress. The other teams could also utilize the same layout, provided their program called for the same V-6. However, as each team's requirements were different, approval of it for team 2, 3, and 4 seemed unlikely.

As the six-week deadline approached, the teams completed their final design iterations, prior to program approval. Cost estimates for tooling, assembly equipment, optional accessories, along with the variable profit calculations were being prepared for presentation to the board of directors. The designs were sufficiently complete to show the overall configuration, utilizing computer-aided renderings of exterior and interior arrangements, underhood and underbody layouts, and clay models of the entire vehicles, along with wood and plastic mockups of the interiors, engine, and cargo compartments. Also included were artists' sketches of scenes involving the different vehicle concepts with potential customers, for exploring potential advertising themes. All of these efforts were the result of a large number of people working virtually nonstop in order to complete their programs in the allotted time. Some workers never left the facility except to obtain fresh clothes. The company kept the in-house cafeteria open 24 hours a day, along with designated rooms at local hotels for those who lived more than one hour away from work.

On the morning of the presentation, the CEO was given a private tour and briefing of the exhibits. He had been monitoring their progress but had not visited the design studios during the work period so as not to expose the teams to any additional stress or influence. During this briefing he had several of his own concerns but he kept silent, other than nodding occasionally to acknowledge the information. Around 9:30, he returned to his office, briefly scanned his desk for new phone messages, then proceeded to the main lobby to greet the board members, most of whom were always early for their meetings. True to form, several were standing around, looking at their watches and frowning. The board's composition reflected the typical distribution: CEOs from other manufacturing firms, investment bankers, retired university regents, and government officials. National's CEO greeted each board member warmly and led them the design studio, located behind the headquarters building.

The four concept vehicle displays were arranged in a large circular format, with temporary walls set up between them. The walls portrayed the detail layouts and artists' sketches of optional exterior and interior trim packages, color coordination, and advertising themes. The first stop was team 1, project C-147, the wagon version of the Centaur, nominally called "Centurion." The first impression of the clay model was that it was a completely different vehicle. Ground clearance had been substantially increased, and more aggressive tires were visible underneath wheelhouse flares. Decorative stripes ran along, up, and down the sides and around the liftgate. A profile of a Roman soldier was highlighted on each quarter panel. The roof rack held water skis, a surfboard, and other associated gear. Scattered around the car were camping equipment and supplies. The board members filed past slowly, stopping to point out salient features, smiling at some of the team members, who had been recruited as actors for the scene.

The same events were repeated at each of the remaining projects. D-561, the sport utility derived from the Orion sedan, appeared with a boxy front-end, chrome-plated grille, large off-road tires, and brush guards around the entire front. A small roof rack contained shovels and other hand digging tools for archeological exploration. A fake "dig" had been staged, with scientists in khaki uniforms supervising "locals" carrying baskets of sand and dirt and screening the contents for artifacts. Decals on the side of the vehicle described the name and funding responsible for the expedition.

The third project, T-205, a midsized pickup, shared the front-end, suspension, and power train with the sport utility, D-561. The scene portrayed was an oil well drilling operation. The truck was caked with mud, loaded down with equipment, and set up as a mobile office, complete with portable computer, facsimile machine, and plotter behind the front seat. Seismic charts, drill bits and lengths of pipe were lying about. Three burly, sweating workers were poring over the plotter as it produced a new chart covered with colored lines and sworls. "Piped in" noise from large diesel engines and other construction sounds completed the image.

The last project, the UAV-100, had very little in the way of background scenery or decoration. The display stood alone, in the center of the area, complete and unadorned. The lights were turned down; some board members questioned whether the display was even open or ready. The CEO told them to be patient and wait. Suddenly, a man dressed in a suit came out around the partition, got in on the driver's side, and pretended to start the vehicle. The screen lights came up and a driving background commenced to move, in conjunction with highway sounds, which now emanated from the screen. Two other people appeared, the vehicle (actually, the background) stopped, they entered it, and it started moving again. After a short time, the background stopped, the background darkened, the people got out and left, and others dressed in casual clothes came out. They removed the rear seats and installed the roof rack pod, loading the roof rack with soft luggage bags and putting boxes of books and records into the rear cargo area. The people then entered the vehicle and pretended to drive away. Then, after arriving at a garage sale, others came out, greeted the occupants, helped to unload the contents, and removed the roof pod and rear roof module and drove away. Other people dressed in swimwear came out and loaded the surfboards they had been carrying into the back, got into the vehicle, and "motored" toward the beach. In quick succession, similar scenarios were played out including a lumber-carrying flatbed, side panels with toolboxes for telephone line repair, and a shallow plastic dump box and liquid tank for spraying vegetables in a garden. All the while, the vehicle's automatic variable height, wheelbase, and attitude were constantly being readjusted for each new load condition.

The directors stood motionless for a while, dumbfounded by the constant ebb and flow of activity, musical tunes, and material fast accumulating next to the set. When the action finally ceased, a very large quantity of equipment, supplies, and body panels was stacked and sorted. The directors all clapped vigorously and departed, talking with animated gestures and exclaiming loudly about their favorite episodes, as though leaving a theater after an enjoyable movie. They went directly to the executive dining room, where discussions continued over lunch. The strategic planning committee was scheduled to meet at 1:30 P.M..

11.4   STRATEGIC PLANNING AND INITIAL DECISION PROCESS

After lunch, the board of directors was joined by all of the vice presidents, assembled in the boardroom. However, for this meeting, the curtains were drawn back and natural sunlight filled the room. An electronic noteboard was arranged opposite the head end of the table. The company's secretary was writing an agenda on the board as the attendees filed in. National's CEO came in last, shut the double doors, and took his seat. "Ladies and gentlemen," he began, "today's meeting has one main topic, to decide on whether to pursue each of the new product programs, which you have already reviewed this morning. Before we begin the formal discussion, are there any general comments?"

One of the senior members, an investment banker, raised his hand and stood. "Mr. Chairman, you have my complements on the superb exposition we viewed this morning. I believe that I can speak for all of us in saying that we have never attended such a well-executed and elaborate show, which we all thoroughly enjoyed. Thank you very much. However, I must add that we didn't see much in the way of new technology, either for manufacturing or environmental improvement. Are you saving these areas for later, or did we miss something in the presentation?"

The CEO paused for a moment, then asked the Marketing Vice President to address the question. The VP said, "Yes sir, we anticipated this question. We believe that, with a little more development of some new catalyst technology, all of our power plants will meet the EPA's definition for ultralow emission vehicle specifications. New manufacturing technology consists of large cavity molding capabilities for the UAV-100 body panels. I will refer to the overhead screen, which will demonstrate this process." He flipped open a panel on the wall and turned down the lights and venetian blinds simultaneously. A large video screen descended, and a computerized animation of the molding process began. The narration described how the plastic material was selected for specific properties, including temperature stability, resistance to crazing, warping, ultraviolet radiation, and moldability. The video continued with the molten material flowing into the mold, pressure increasing as the molding screws moved in, and finally the molded parts exiting the mold. Next the four pieces constituting the UAV-100 front body were adhesively bonded together and then welded to the front floor pan and chassis rails. The result was a robust subassembly. The video stopped, and the lights and window shades came up.

The marketing VP said, "We still need to run durability tests on the plastic and steel assembly, but we think it will pass without any problems. The new catalyst material is a bit more problematic. The development testing is well underway, but some of the chemical reactions are not being treated, and there are several byproducts that need to be addressed. The supplier is confident that these problems can be solved quickly. However, we are exploring another new technology as a backup."

Another board member, an officer in a chemical company, wanted to know the name of the new catalyst supplier and if his company could get involved in the development. After supplying this information, the secretary then distributed the agenda, and the committee reviewed it for a few minutes.

The first half dozen items concerned expenditures for building renovations for next year's new models, in the factory, test labs, test track, and rail head distribution center, for new lighting, concrete work, fixtures, and new assembly machinery. With only a moderate amount of discussion, all budgeted items passed unanimously. Also approved was the compensation plan for all nonunion employees based on the contract terms, which still had two years to run.

The last item was the evaluation of the four projects beginning with the C-147. This project represented the least amount of investment for a new model. One of the other manufacturing board members asked about the incremental market potential. The marketing VP responded with a projection of 15,000 maximum additional annual units. Production timing was estimated at 10 months from design freeze, which typically occurred 90 days after approval. Incremental variable profit was estimated to be approximately $2250/unit for the Centurion, compared with $2000/unit for the Centaur. Tooling and other incremental expenses totaled approximately $4,750,000, so it would take a little over one year to pay off. The controller questioned whether the value of this project could be determined without knowing how it compared with the others. Other members acquiesced and the C-147 was tabled. None of the team 1 members were in attendance to object, although the vice president of engineering could have rebutted if he had chosen to. The main expenses were for new tooling for the new rear drive axle, reshaped quarter panels, and the longer roof for the wagon.

Projects D-561 and T-205 were considered together, since the major investment would affect both vehicles. Despite the fact that two new vehicles could be marketed for the price of one, there were reservations concerning crashworthiness and fuel economy for both models. The engineering VP did discuss their benchmarking efforts on competitive models. He told the group that the test lab had been experimenting with some new frame horn configurations which, based on static and drop-tower results on prototype hand-built parts, looked promising. Dual air bag deployments resulted in acceptable head-injury numbers, but front floor pan intrusion was unacceptable for the 40% offset 40 mph impact. This test was being promoted by the insurance industry, and very few current production vehicles had good results from it. Fuel efficiency improvements were also being studied, utilizing competitive vehicles, but modified for lower power outputs and emissions. Horsepower levels were deemed acceptable except for four-wheel driving modes, both on- and off-road. Supercharging the V-6 engine produced more power but also more stress on piston wrist pins, connecting rod and crankshaft bearings, some of which had failed under durability testing. Further developments such as larger bearings and journals were under study.

A V-8 engine, with substantially more torque, was desired but unavailable due to limited space for an additional transfer line in the engine plant and the lack of a more durable four-speed overdrive automatic transmission. Fuel economy and emissions would probably have been slightly worse, due to a 75 lb weight penalty. In the end, the board approved development of the V-8 and bigger automatic transmission for later introduction on the D-561 and T-205.

The UAV-100 presented an entirely different set of issues to the board. One of the academic members raised a question concerning global warming and fossil fuel usage by the new vehicles. He said that the newest program from National should be setting a precedent for minimum environmental impact, regardless of the minimal demand or lack of profitability. He further added that this proposal fit perfectly into National's recently revised mission statement, to be the most innovative, conscientious, and forward-thinking automobile manufacturer in the world, with the most advanced research and development operations, creative design bureaus, and state-of-the-art manufacturing facilities anywhere. He submitted that UAV-100 did not go far enough to satisfy the corporate objectives and should be redesigned as soon as possible, to fulfill the goals outlined in the mission statement. Several other board members added similar comments, while corporate officers challenged the board, defending the current scope of the UAV-100 program on an affordability basis and attacking the lack of a viable market and the excessive development and production costs of the academic professor's proposal.

The CEO cut short the increasingly rancorous meeting, then said, "Ladies and gentlemen, please! I know you all have our best interests in mind, but I need to bring us back to reality. The truth is we have both short-term and long-term problems. We see the C-147 and D-561/T-205 as near-term solutions, while the UAV-100 represents our best guess as to what will be needed in three to five years. We need to maintain market enthusiasm and buyer interest with these new, exciting, but frankly interim, products, while we pursue the long-term objectives as the most advanced car manufacturer. Please now, help us to figure out how we can meet both of our goals.

I can tell you that, as the current year projects out, we will probably lose approximately $47.5 million by year end unless we can find a way to sell our cars. This means that an additional 40,000 total units, in the proportions of roughly 2:1:1 for the Centaur, Orion, and Pollux, need to be sold in the next five months, in order for us to just break even. If you can help with contacts at the major leasing, car rental, and fleet companies, please let the secretary know and we will be eternally grateful. We can give them a very good price for quantity purchases of these exceptional cars. Concerning our next project, we will continue our discussion after the afternoon break. Thank you for your attention."

The meeting adjourned for an hour to permit phone calls and other business to be handled. Some of the board members huddled in one end of the room while the VPs dashed out to return calls to vendors and project coordinators. After the break, the marketing VP received a call from one of the largest megadealers in North America, expressing interest in a volume purchase of Centaurs. Another call came from an electric utility in the Northeast looking for a price on cars for meter readers, this time a referral from the board member in that business.

The meeting reconvened with the continued discussion on the costs associated with developing each project. Table 11.1 shows the distribution of costs for purchased components, fabricated subassemblies, final assembly labor, materials, and so on, for the four vehicles, along with design and engineering, testing, and other product development expenses, and wages and salaries. Also shown are the competitive vehicles that were benchmarked for the study.

In the end, after analyzing the cost factors, the decision was made to proceed with each project, since retained earnings set aside for new product development remained nearly constant, (based on sales and profit projections), and the time available to complete the new programs, resulted in their completion in each of the four succeeding years: C-147 after one year, T-205/D-561 in two to three years, and UAV-100 in four. This last outcome assumed that the new technology for an alternative hybrid engine/drive system and fuel cell would be perfected in that time.

11.5   MARKET RESEARCH AND IDEA GENERATION

The board's decision meant only that the projects could move to the next phase of the process—i.e., market research, to confirm the direction taken by the board, and to modify, expand, or reduce any program that did not meet the objectives agreed to in the decision. It also gave the researchers an opportunity to test several hypothetical scenarios pertaining to the new technology planned for introduction in the UAV-100. These included a virtual driving simulation of the vehicle, under a variety of operating conditions (e.g., intercity traffic, expressway, off-road), inclement weather, maximum gross weight, and so forth. The research clinics also repeated the same presentations that were made to the board of directors, complete with the same employee actors, music, and sound effects. It was to be a new experience for many research clinic recipients.

The same number of people had been selected from previous clinics and utilized several times. The research department had been contemplating a different approach, particularly, but not exclusively, for the UAV-100. It was felt that a new selection procedure and criterion were needed, to ensure that an accurate sample of responses were collected. A broader cross section of the population would provide a more representative sample of the average vehicle user and allow a further analysis of his or her concerns with the new automotive technology to be contained in the UAV-100. The main problem was one of logistics: trying to get a traveling "road" show around the country, collect all the data, analyze and interpret the results, and get recommendations to the new product committee in time to be enacted into the program.

One theory currently in vogue allowed that random sampling produced random results, which could have explained the high probability that previous data was misinterpreted or misused. National had been subscribing to a market research service offering assurances that their system of picking candidates from mailing list recipients of automotive publications was the most accurate method for determining demand for a new product. Theoretically, car enthusiasts are a key buyer group if there are sufficient numbers. But lately, the research department was beginning to have doubts about the credibility of this method. National also relied on its own sampling methods, relying on direct contact with previous buyers and dealer surveys to supplement outside data. For the four new projects, something completely different was in order.

The foremost method utilized to collect new product ideas is the brainstorming session, where selected participants are exposed to stimuli on specific new product themes and are requested to expand or elaborate on them, either with or without constraints. For success, two factors need to be regulated, the criteria for selecting participants, and the number and complexity of the constraints. The research department had determined a number of selection criteria that they desired to test. These included specific vehicle make and model ownership, geographic and demographic data, and previous vehicle ownership trends. They surmised that former and current utility vehicle owners might be the best prospects, not only to fine-tune the details for the C-147, D-561, and T-205, but also develop the characteristics of the UAV-100. In addition, interests in bicycling and gardening were chosen as potential buyer identifiers for the UAV-100. The research contractor had provided data on current owners of minivans, small pickups, and sport utilities for the market research scientists at National to study.

The data on current utility vehicle owners showed very little in the way of fresh research. Owners liked certain characteristics on each vehicle, such as a commanding view of the road and utility space, but disliked the poor fuel efficiency and harsh ride characteristics. They were little help in learning how the vehicles could be improved, other than minor modifications. Buyer demographics could not be relied on to sort the key characteristics, identified earlier as potential buyer indicators. Therefore, a series of market research clinics was created around several key buyer groups, with the hope of discovering some new, breakthrough findings that would permit further differentiation and analysis. The marketing vice president reviewed the objectives with the teams. The clinics were organized around the following well-established idea generation techniques, to be introduced and controlled by trained facilitators:

- *Attribute Analysis*: List all the attributes that can be noted, and look for combinations, adaptations, or modifications for potential improvement.

- *Conjoint Analysis*: Looking for the interrelationships between several dependent variables that might all be affecting the outcomes of the research.

- *Forced Relationships*: Compare and contrast the above list for new relationships between them that may not have been apparent before. Try reversing cause and effect, interchanging or transposing components, or combining and separating functions and operations.

- *Morphological Analysis*: Rank-order the most important attributes to find new relationships between them. Finding the most important component in a system could lead to other subsystems, new components, and so forth.

- *Brainstorming*: This is one of the most universal methods that can produce many new ideas, if properly channeled. Some of the guidelines are as follows:

    No criticism or evaluation: This is the screening process, the next phase in the process.
    No limit on type or outcome: Anything and everything goes.

Collect as many new ideas as possible within a given time frame, one to two hours maximum. Encourage combinations and improvements by building on each new suggestion, as they occur.

The clinic candidate selection process was developed next. The same database of current minivan, sport utility, and small pickup truck owners was utilized as the selection source. A random selection program picked 1050 potential candidates for final screening. Letters were prepared and mailed to the recipients, requesting a response as to their interest in participating in the clinic. They were also asked to rank their interests in a number of specific hobbies and activities. Next the returns were screened by occupation, interest level, availability over certain time slots, activities, and hobbies. Then a final selection based on a variety of factors was distilled down to the list of participants in five geographic areas. The clinics were planned to take place over a five-week period, with analysis and recommendations to be completed within two months after startup. Properties similar to those utilized in the board presentation were prepared for the clinics. It was planned to transform the properties into hands-on components that participants could pick up, handle, assemble, and disassemble, either when single or as a group exercise. Facilitators would be standing close by to assist them if necessary. Before the hands-on session, the participants were to hear a brief presentation on the goals of the program. Following a question-and-answer period, the facilitator would then engage the group in the various exercises listed above, stopping only to allow them to reconnect, if one of the group diverged from the group on a tangential issue. For example, it was anticipated that wheelchair accessibility would have to be addressed, even though the vehicle configuration probably did not readily lend itself to that capability. The entire session was to be recorded on videotape so that useful ideas could be collected without disturbing the thought process.

Some of the results of the first few clinics confirmed the validity of ideas collected at earlier sessions. Several concepts were "proved out" through the hands-on experience, including the ease of mounting and demounting the roof cargo pod and the installation of rear seat modules. One idea that changed as a result of the clinic was the replacement of the removable back panel, behind the front seats, with pocket doors that opened by sliding into the C-pillars. This was deemed necessary to eliminate exhaust gas buildup inside the passenger compartment when driving without the rear roof module. The original concept called for a completely removable panel that the owner could put back in with aircraft-type quick-disconnect fasteners and special gaskets, to form the seal around the perimeter floor tracks. Several alternatives came out of the hands-on session, including a hinged-door bulkhead, a fold-down rear window, and sliding panels that would disappear into the vehicle side panels through the utilization of a track system. All of the new ideas were collected to be evaluated later. The clinics also confirmed trim package and color combination preferences. Various interior seating options were tested, as were three configurations of instrument panel gauge layouts, control knob styles, and rear view mirror sizes and locations. When overwhelming preferences were expressed for a given set of conditions, in week after week, several new variations were introduced to measure their effectiveness against changing conditions. It was found that the new midpoint entries had inconsistent results compared with the original set of variations, and the practice was discontinued.

11.6   SCREENING AND EVALUATION

The screening process had evolved into a checklist-based review of new product ideas, according to three main criteria; marketability, manufacturability, and profitability. Since all three parameters had to be met with at least minimum performance, very few new ideas were strong enough to withstand all three criteria. The marketing criteria was based on the results of the surveys conducted in the above clinics, on competitive intelligence collected and analyzed by the marketing staff, and on the statistical models of future customer demographic profiles, buying habits, and economic trends.

The manufacturability criteria were based on a jury of experts, all specialists in various aspects of component fabrication, welding, stamping, casting, and assembly, who reviewed each new design and concept. While their primary concern was being able to build the vehicle according to the design, they also brought up other issues such as warranty problems, serviceability, and repairability, based on past job assignments. The financial criteria concentrated on the credibility of the profitability predictions and a list of potential "what if" scenarios that could change the outcomes of the projected profits. Some of the more significant scenarios included price increases by suppliers, union strikes, and a fire in a supplier's warehouse. All of the aforementioned items are shown on the checklists in the Appendix.

The only project that did not survive the screening was the hybrid power plant for the UAV-100. Based on the lack of a viable market, the program was to continue development utilizing a conventional engine and transmission. However, some effort was directed to maintain a hybrid presence in the program and wherever possible, provide input and design assistance, so that the hybrid could be incorporated with a minimum of extra design work.

11.7   PRODUCT PLANNING

The next phase in the process called for the product planning department to develop and issue, for approval, new vehicle program proposals for each new vehicle being considered. The proposal form, as it appears in the Appendix, covers any unique market, financial, engineering, manufacturing, service, and other issues based on the information available, along with the target market or benchmark competitive vehicles, except in the case of the UAV-100, which was without any current competition. The purpose of the proposal was to gain a consensus of all affected participants as to the viability of each program and to gain commitments to meet the schedule for production. If there were problems with any requirement, the concerns were to be presented at the next new product review, typically held the first Monday afternoon of each week. (*Note*: specific details have been omitted from the proposals on engineering and manufacturing issues, which would have been identified later in the project.)

The new vehicle proposals had been distributed one week before, and now the agenda for the new product committee meeting was overflowing with requests, from virtually all participants, for time to address their concerns. The schedule was divided into 10-minute increments, which, while not allowing much time for elaborate presentations, forced the speaker to be concise and unemotional. There were many concerns about meeting the weight targets, the corporate average fuel economy, and crash performance criteria for each program. In addition to the aforementioned worry over the fatigue strength of the supercharged V-6 engine, the gross axle weight ratings for the D-561 and T-205 did not seem to address the additional loads from trailer towing. These vehicles would have to be restricted to a tongue weight of 100 lb and a trailer weight of 1000 lb, maximum. These limits made the vehicles marginally competitive with their benchmarks, and definitely noncompetitive compared with several other competitors. Even if the rear axle rating, which includes springs, tires, and wheels, could be raised sufficiently, the engine horsepower would be barely sufficient to propel the vehicle by itself, let alone with a trailer. Supercharging was meant to be a short duration event to supplement power during passing or other limited operations. (The proposal forms for the four projects have been included at the end of the chapter).

In the end, all of the first three program management teams expressed the same concerns about meeting the targets for mass, driveline durability, and crashworthiness. There was no reason or argument presented that was going to stop or slow down any of the developments. It only remained to monitor their progress and apply increased resources when crises occurred. A totally different situation was presented by the UAV-100 program manager. Defining the worst-case scenario was a major milestone, as were the durability cycle and crash test criteria. Proper operation of the wheelbase extender and locking mechanisms for removable panels was also deemed critical. However, at no time did the team express anything but optimism toward meeting the goals.

11.8   DESIGN AND ENGINEERING

All four new product proposals were unanimously approved, and design release schedules were established. Due to production timing constraints, the C-147 was given first priority, the D-561 second, the T-205 third, and the UAV-100 fourth, although some engineering work would be done simultaneously, depending on the system or subsystem priority. Dual releasing of common components for the D-561 and T-205 was also permitted.

Procurement was allowed to discuss all four programs with selected suppliers so that their input could be cultivated and maximized more effectively. Vendors were required to support development of key components and subsystems, through their own in-house testing programs, those in other laboratories, and in prototype vehicle road testing. It was now all up to the engineering department to coordinate efforts and make sure it all came together right and at the right time. Weekly new product meeting schedules were set up so that key personnel could attend without any overlap and without causing burnout. These mini-design reviews were intended to correct any communication or data transmission problems between handoff groups. Usually the agenda contained only a few items, which were discussed quickly. If an issue could not be resolved in the weekly meeting, it was referred back to a study committee for priority assignment, and a resolution in keeping with the overall master program timing chart was scheduled.

11.8.1   Advanced Design Simulations

While the program teams were being organized, initial designs were being scrutinized for any major problems encountered prior to the detailed engineering tasks were assigned. This effort served to identify a workload that had not been previously found and to address it as early as possible, as part of the Simultaneous Engineering philosophy.

Various simulation programs were also initiated along with the design/release activities. These included advanced suspension and driveline emissions durability, and front and side crashworthiness and rollover survivability modeling based on new worst-case scenarios from the research department. Other programs included maximum optional equipment gross vehicle weight performance testing, which consisted of a combination of 0–60 mph acceleration, 60–0 mph braking, and lateral acceleration around a 100-foot diameter circle, relative to maximum-loaded benchmarked competitors. Later, after initial design studies resulted in mass and cost estimates for the new programs, simulations of warranty and collision repair cost, quality of manufacturing processes and functional operations were planned for completion. Results of the mass and cost simulations and the combined optimization effort are shown in Tables 11.1 and 11.2 at the end of the chapter. Before analyzing those results, the initial design activities were begun under the following conditions. First, the competitive benchmarked data on cost, mass, dimensions, and other specifications were fed into the master computer. Next, the changes to generate the new designs were inputted for comparison. The new design was constructed by comparing new digitized requirements to the existing benchmarks and manually overriding each specification until the new design was complete. Then a graphical record was constructed so that the new design could be directly measured to the benchmark or older design, in order to identify any discrepancies and correct them. For example, the simulated total mass of the initial D-561 design had the wrong size and mass for the power train, which showed up as a conflict in the computerized layout and was later corrected.

After the initial designs were accepted by engineering, the various department heads were asked to compile lists of tasks requiring new manufacturing or process technology for production. Potential end item suppliers were also selected, along with any second- or third-tier subcontractors who were promoting new ideas. Procurement had been funneling these new as well as established vendors to the engineering groups for several months, with the prospect of front-loading any problems that would require a significant increase in development time over the normal schedule.

Program timing estimates were developed based on earlier experience, but tempered with the knowledge of new Simultaneous Engineering techniques. One of the keys to guaranteeing success with these new techniques was making sure that all suppliers had instantaneous access to the level of information they needed. Linking of computer systems, providing the basic access codes, and software compatibility were followed with joint meetings where the four programs were presented in sufficient detail to enable each supplier to work at a pace that was within its capabilities, with lean resources, yet stretched to ensure compliance with all parameters.

The first simulations involved crashworthiness comparisons with current safety standards, as well as anticipated new ones. The current front crash test involving passive restraints, MVSS 208, had not been a serious impediment for the last three model years, even when the New Car Assessment Program (NCAP) tests were run at 35 mph instead of at 30. However, the proposed insurance industry–sponsored deformable 40% offset barrier test at 40 mph presented a significant challenge, to be anticipated, even though the National Highway Traffic Safety Administration (NHTSA) had not yet committed to supporting it.

Another crash simulation under development, in which innovation played a major role, was vehicle rollover, where a multitude of injury mechanisms and the lack of a repeatable test had stymied designers for many years. Restraint systems engineers could modify existing seat belt systems to protect against side window ejection, provided the rollover was one complete revolution. When the vehicle rolled over more than one time, occupant dynamics could change significantly due to increasing vehicle deformation, changes in vehicle angular acceleration, and interaction between the occupant and the vehicle.

The passive restraint design group was evaluating selected potential vendors to develop remote sensing for out-of-position occupants, as the first stage in determining whether air bags or other restraint systems could be deployed effectively in a rollover scenario. Small charge-coupled device cameras were mounted in the headliner of a vehicle, prior to running a rollover test. Results were mixed due to the deformation of the vehicle roof, which caused the camera to go out of focus and lose power temporarily, producing unacceptable pictures. However, the clear pictures did provide some new ideas on how to contain occupants, particularly in the rear seats of the sport utility, where limited head room helped to control movement.

New exhaust emission control requirements were anticipated to be implemented by 2000, for the 2004 model year, with carbon dioxide control for global warming in addition to the other gases. The nitrogen oxides presented several new problems that the engine design group thought could only be handled with the addition of a new, largely experimental reduction catalyst. All new engines were mandated to meet these new regulations even if the EPA backed off or delayed its plans. It was seen by senior management as a marketing advantage if all new National models could meet the ultralow emission limits. Therefore, all new programs inherited this mandate, as well as carryover products with new engines.

The new V-6 supercharged engine was viewed as having the best chance of meeting the new standards. This program had been in development for several years, with various configurations installed in prototypes and running in high-mileage durability cycles, typically now, at 100,000 miles, without a major failure. Emission test results were acceptable without the supercharger operating, but only marginal when it was running. There was also a conflict between the supercharger and the variable valve timing module, between idle and 2750 rpm. Tight engine compartment packaging resulted in lengthy and constricted routing of the intake air plenum, and air cleaner back pressure caused a lag in supply to the supercharger. The engine controller could not consistently provide the proper ignition advance to match both systems since the requirements appeared to be in opposition. The controller software would have to be modified to reduce the valve timing advance until after the manifold pressure boost stabilized itself; otherwise, a large hydrocarbon spike developed during the transition stage, not an optimum solution. Alternatives were being explored.

Fuel efficiency also suffered somewhat during supercharger operation. When the prototype derived from the Achilles benchmark was set up with an experimental V-6 engine and ballasted to maximum gross weight, fuel economy dropped more than 27% from nominal emissions test weight and nonsupercharged operation. In case some other mass reductions could not be implemented, supercharger usage restrictions were being considered. This development would be a loss from a marketing perspective, since both trailer towing and off-road capabilities would be adversely affected, as well as certain performance parameters, such as acceleration and braking.

11.8.2 Simultaneous Engineering Examples

The following process description comes the closest to the classic definition of Simultaneous Engineering, where every activity is planned, coordinated, and directed to eliminate wasted time and motion of every phase in new product development. Many tools and techniques have been introduced over the last several years to provide engineers and designers with the means to achieve their goals, i.e., optimization of manufacturing cost, minimum mass, minimum warranty cost, continuously improving quality, meeting market requirements, meeting or surpassing functional operation targets, among others. Performing the tasks correctly, the first time, without any rework, redesigns, mistakes, changes in direction, and so on, is another equally important goal. The philosophy is known by many names, including concurrent engineering, customer-focused quality, quality function deployment, cross-functional platform teams, among others. It suffices to say that any well-thought-out, coordinated effort to get a new product to market faster and more cost-effectively than before will be worth its cost in any monetary term, regardless of the name of the program.

The day after program approval by the board of directors, the functional engineering department managers met with the four program managers to plan the various work tasks; organize engineering specialists and component group, cross-functional, and liaison teams; and assemble time lines for each major workload item. Some tasks were to be shared with selected suppliers who possessed the same computer design software packages, to the extent that the machine interfaces were virtually seamless. Others were to require a new system of devising alternative solutions to heretofore unexplored areas, particularly for the UAV-100 program, with its unique approach to manufacturing and unusual utilization of new materials. The new program teams were established in separate quarters, along with manufacturing engineering, service engineering, and marketing personnel. Suppliers of major subassemblies and systems also participated in the team meetings and discussions prior to decisions. Tier two suppliers were kept informed of decisions affecting their components by a special team representative from procurement.

Weekly timing and decision meetings were held each week, usually on Monday mornings, unless other corporate functions took precedence. Representatives from each program area rotated as spokesman for a particular week. Meeting minutes were recorded for future reference and to allow the timing subcommittee to update the Gantt charts after the meeting. Senior management appointed liaisons to the program teams, and an officer attended every meeting possible, depending on the level of other problems at the moment.

The component group teams consisted of engine, transmission/final drive, body, chassis, and electrical and mechanical systems. Representatives from each of the component release groups, mechanical engineering, manufacturing engineering, reliability engineering, major suppliers, system development engineering met every Tuesday morning, after reporting to the new product committee on Monday. The same CAD software was distributed to all of the designers, release engineers, specialists, and suppliers representatives so that any data, dimensions, and specifications could be displayed such that all affected could review any new information or decisions that affected individual components, subsystems, systems, or the entire vehicle program.

The engine team consisted of release engineers and designers for components, including the block, cylinder head, crankshaft, camshaft, pistons, and connecting rods. Supplier application engineers covered their components such as piston rings, bearings seals, fasteners, and gasket materials, which were both hard composites as well as foam-in-place gasket materials. The computer was programmed to resolve any interference and fit problems that would have been encountered in the assembly process, utilizing a form of dimensional variation simulation, performed during a trial assembly. Gauges and fixtures were checked at the same time. Clearance issues were addressed by the reliability engineers, who analyzed each instance and decided if the situation required a change.[1]

One of the main jobs of the teams was to plan and execute the major improvements decided on by the new product committees, for each program. For the four programs, the tasks ranged from updating mass, manufacturing cost, and warranty reduction targets, to increasing power-train durability targets with the all-wheel-drive fluid couplings, and maximizing engine horsepower, utilizing variable valve timing without sacrificing emissions, fuel economy, or durability.

Tasks unique to the C-147 program included front crash performance, with the new larger engine and added mass from the rear driveshaft, axle and rear body additions. The new engine also had to pass the 50,000-mile emission durability test, which fortunately could be performed on a chassis dynamometer, utilizing a modified Centaur vehicle, ballasted to the new C-147 projected mass. Maximizing engine horsepower involved selecting several iterations of cylinder bores, piston stroke length, and piston mass. The Design of Experiments concept was utilized to provide optional configurations. From the database that was developed when the Centaur 2.0-liter engine was created, several iterations were calculated for a displacement of 2.5 liters, the size thought to provide adequate power while incurring minimum weight, emissions, and fuel efficiency values.[2]

The tasks for the D-561 and T-205 programs also included mass, manufacturing cost, and warranty targets. However, the major effort was directed at the frame and suspension, since National had not yet produced a separate frame vehicle. Not all of the engineering staff were convinced that a separate frame was the right decision. Many thought that there were more cost-effective alternatives to a separate frame, including the following: unitized body with pliable control arm bushings, to control NVH, sound deadening to counteract noise, additional underbody cross members to increase

[1] Paul I. Hsieh, Robert Eugene Lee, and Brian L. Torma, (Chrysler Corporation), "Combining Reliability Tools and Simultaneous Engineering," *Automotive Engineering International*, 106, no. 8, Warrendale, Penn. (August 1998), 66–68.

[2] W.R. Carey, (Eaton Corporation), *Tools for Today's Engineer, Strategies for Achieving Engineering Excellence*, sec. 6, "Design of Experiments," SP-913 (Society of Automotive Engineers, Warrendale, Penn., 1992) pp. 43–52.

body stiffness, torsion bars tuned to limit body roll, and an air bag suspension to provide adequate wheel travel with minimum body movement. The problem with all of the above was the limitations put on payload due to additional mass.

Another challenge was the front crash response and subsequent repairability of the front frame horns. These extensions of the frame side rails were designed to collapse at a specific deceleration rate, which then signaled the air bag sensor that deployment was required. This threshold speed, typically between 12 and 15 mph, on a barrier equivalent basis, was specified to protect unbelted large males from serious injury in a flat frontal 0 ± 30 degree impact. The problem was not the air bag performance or the frame horn collapse. The insurance industry was threatening to surcharge any frame-type vehicle that required full frame replacement from any low-speed impact, resulting in nonrepairable frame horns. This action could render the D-561 and T-205 noncompetitive for insurance cost, one of several key parameters in the cost of ownership. Other manufacturers had unilaterally concluded that frame horn replacement was too risky since the repair industry was not regulated, and lawsuits had begun to appear on this issue as a result of accidents and injuries from improper repairs. The chassis teams from both programs had decided to formulate a joint business case analysis to address this potential problem with a proposal to test an alternative repair procedure, in conjunction with research centers connected with the insurance industry. This was based on the premise that the liability exposure from an improperly replaced frame—i.e., a mistake made in disassembling or reassembling a mechanical or electrical system—was the same as an improperly welded frame horn. It was thought that the section repair could be simulated by initially comparing an undamaged frame with a sectioned one in the impact analysis program, followed up later with fatigue and impact tests on repaired frames.

Another challenge for the D-561 and T-205 engine teams was the plan to include both a supercharger and variable valve timing on the 3.0-liter V-6 engine. Neither of these were familiar to National engineers except through tear-downs and ride-drive comparisons with competitive vehicles. Fortunately, one of the potential suppliers to the engine for the Achilles was marketing a supercharger similar to that on the Achilles. The only problem was that the suppliers design did not pass the Achilles manufacturer's engine durability test and was rejected. The reason for the rejection was due to a poorly designed air pressure boost regulator, which caused overboost in the cylinders, which in turn caused knock and added stress on connecting rod bearings. A new design seemingly corrected this deficiency and awaited testing in a production engine. The team received approval to purchase an additional Achilles model, replace the production supercharger with the redesigned one, and subject the vehicle to the engine durability cycle.

The variable valve timing issue was intrinsically more involved since the 3.0-liter V-6 engine was a single overhead cam configuration (SOHC) and needed to be a dual overhead cam (DOHC) in order for the timing mechanism to work properly. This meant a new cylinder head design, and a new valve location and valve spring arrangement. Normally, this process would have required several months for a layout, experimental castings, machining, and prototype engine buildup and dynomometer testing, to prove out the design. Fortunately, a DOHC design had already been completed as part of an upgrade study and only needed the fit-up of the hydraulic metering system. Competitive system tear-down results were available, and a potential supplier was interested in the job. However, the competitive system was patented, which meant that a royalty would have to be paid for every one sold, unless major modifications or sufficient improvements could be designed into the new concept, to preclude any infringement. The supplier representative thought that their engineers had come up with several suggestions and were ready to try them out. A purchase order was cut to permit the supplier to submit a prototype unit within 30 days.

The UAV-100 program manager had been meeting with his Simultaneous Engineering (SE) teams and was listing the major design and development tasks, for timing and workload balancing purposes. Some of the more important jobs were as follows:

- Feasibility of the plastic composite materials to withstand accelerated durability
- Feasibility of movable rear body panels and axle concept
- Certification of 2.5-liter four-cylinder engine
- Feasibility of a hybrid power train
- Feasibility of air over hydraulic suspension system for entire gross vehicle weight (GVW) range
- Determine merits of aluminum versus steel for body construction

The program director selected leaders for each of the major jobs and met with them as a group. The workload plan was established based on previous research and development performed by certain suppliers, through actual accelerated durability and fatigue tests on component parts and subassemblies. The four-cylinder engine was deemed to be the easiest due to the previous effort expended by the C-147 program.

Durability tests on the C-147 project were progressing nicely, but the computer model for predicting fatigue life was not yet up to specifications. Unusual stresses were detected in the middle crankshaft journal bearings, which necessitated a wider bearing surface. Noise, vibration, and harshness (NVH) concerns were also addressed by using tuned motor mounts that canceled out the major engine harmonic resonance automatically through a hydraulic memory mechanism. Component fatigue life test results, submitted by suppliers for analysis, showed average wear characteristics for piston outer surfaces, rings, and connecting rod bearings. Combustion residue on the top of the piston indicated that an inadequate swirl of the charge was being imparted into the chamber. This in turn showed that changes in the intake valve were necessary.

Warranty issues from earlier models were being tracked on the new programs to ensure that they were addressed properly. Body opening and closure fit consistency, wind noise, exterior panel and interior trim fit and finish, electronic and electrical system reliability and mechanical component failure trends, were the prime areas of concentration. The key measurement of warranty claim frequency is conditions or occurrences per hundred vehicles. Several designs of experiment projects were organized to ensure that each new product program received the benefits from the previous effort.

Repairability requirements were also being considered for inclusion in the new programs. These included development of repair procedures for all exterior panels regardless of material, partial replacement procedures for all critical structural collision components, refinishing, and identification of virtually all materials with recyclability potential. This last requirement included all metal parts, major exterior and interior plastic components, instrument panel, carpeting, underlayment insulation, seats, and so forth. The position on recyclability resulted in a conflict with purchasing, which was involved in a seat assembly simplification project. The project required a 50% reduction in assembly time, to be achieved by reducing parts and labor. However, this usually meant that subassemblies were combined with others, metal reinforcements were molded inside plastic assemblies, and so on. Special attention was required to ensure that the metal could be easily separated from the plastics without adding costly fasteners or excessive labor. In the case of the instrument panel, the metal substrate was designed to be molded in but partially visible from the back side. Removal could be accomplished by breaking away several heat staked studs, which allowed the substrate to be pulled free. A single plastic material was specified for the instrument panel, even though multiple properties were required. A new design provided for high-impact strength areas by further utilization of heavy steel reinforcements, separate from and behind the instrument panel, to withstand the air bag deployment. This reinforcement, now commonly referred to as a "cross car beam," has other functions related to side-impact protection, body shake, and NVH control.

11.8.3 Design Review Process

The new product committee designated that a regular review of the new designs should be undertaken at least once every two weeks, unless more frequent reviews were necessary. The design review meeting had the following format:

* Restatement of program goals
* Review of previous meeting decisions
* Review of changes and assignments from previous meetings
* Review of the parametric optimization format

11.8.4 Benchmarked Parametric Target Analysis

The last item was the heart of the review process because it showed the latest results of all the various design optimization efforts going on for each program. Large video screens around the conference room displayed computerized images of specific design details, results of fatigue, durability and crash test data, and target proximity of the design parameters identified earlier. These included the following dependent cost parameters: crashworthiness, warranty, functional performance, environmental compliance, durability, damageability/repairability, recyclability, and quality. The independent cost parameters were mass and manufacturing complexity. The parameters were nonlinear quadratic functions comprised of multiple terms, representing all of the variables defined above. For example, functional performance included a term for an acceleration time of 0–60 mph, another for braking, and a third for lateral acceleration. The nonlinear characteristics provided measurements of how close the current values were to the target. At the first meeting, the variances for each new product program ranged from 47.6% for the warranty expense to 5.28% for crashworthiness. The other parametric variances fell in between the above. The eight component groups from the four programs each had eight parametric displays arrayed in large monitors around the conference room. The agenda called for the component group of each program starting with the C-147, to summarize its respective positions separately; afterwards, the new product committee would resolve any conflicts between the groups.

Within the C-147 program, the power-train group displays showed small to moderate challenges to benchmarked targets for the following parameters: functional performance, 5.78%; environmental compliance, 7.43%; warranty (leaks), 14%; durability, 17.6%; quality (NVH), 26.0%; and damageability/repairability, 39.1%. Crashworthiness and recyclability both were under their targets by 9.8% and 3.5%, respectively. The target for unit variable profit was met, and the vehicle mass was within the specification tolerance ($\pm$ 10%) to achieve the emissions, fuel economy, and performance targets, but senior management wanted to come as close as possible without incurring any large capital expenditures.

11.8.5   Program Optimization Studies

The other C-147 component groups reported similar findings, which were also analyzed in a similar fashion. The next step was resolution of any conflict between the groups and consideration for any "what if" scenarios that might become reality in two to three model years. The other three programs repeated the above sequence in turn, with the D-561 and T-205 treated as a combined program. The UAV-100 program took twice as long to present and analyze due to its unique nature, the new materials and manufacturing processes required, and the complexity of the overall design. Cross-program discussions were also in progress all during the program presentations. The four power-train teams were considering the three different technologies for maximizing engine horsepower, specifically, increased displacement for the four-cylinder engine, supercharging for the V-6 engine, and variable valve timing for both engines. The same tests and simulations were performed on the three technologies for each vehicle program application. From an investment viewpoint, it made sense to utilize the same features across the board, if possible, taking advantage of volume efficiencies.

However, constraints were not consistent between the programs. Based on computer simulations, the cylinder wall thickness in the C-147 limited the amount of bore expansion, while the deck height limited the amount of stroke increase available. Overboring could jeopardize cooling efficiency, which could cause block warping, leaking, or severe piston ring wear. Supercharging added higher than acceptable stresses to the crankshaft and main bearing journals that could not be easily remedied in the four-cylinder engine without major upgrading costs. Variable valve timing resulted in reliability and durability problems during the emission cycle test, caused by instability of the valve overlap mechanism. Sizing of the vanes and valves was feasible, and the potential solution of the problem involved virtually no increase in mass.

The Design of Experiments simulation showed that slight changes in each of the valves and vanes could individually net out a small increase in horsepower, but various combinations resulted in a greater increase than any individual action. An array of available cylinder bores, strokes, crankshaft, and main bearing widths and diameters were selected for analysis. Variable valve timing iterations in the form of small damping/accumulator chambers, line routing options, and spring rates were analyzed in a similar manner. This resulted in an optimum set of design constraints for the 2.5 L four-cylinder engine for the C-147 program. The 3.0 L V-6 was also analyzed prior to finalizing its design.

The UAV-100 program faced unusual challenges since there was no clear-cut alternative for the power-train option, (i.e., over the base 2.0 liter I-4 gas engine), except a 1.2 L three-cylinder diesel engine and an 80 hp (horsepower) electric motor with a battery pack that allowed a 100-mile range between charges, assuming the maximum speed is held to 45 mph.

Fuel cell technology was also being considered, with both water- and gasoline-based systems under analysis. Both had advantages and disadvantages in performance, thermal and mechanical efficiency, operational convenience, and safety. In terms of environmental compliance, each system met the EPA's phase II ultralow emission vehicle requirements (not zero emissions, due to excess carbon dioxide from the water system and gasoline vapors from the other system). With either fuel cell (i.e., water- or gasoline-based), the maximum speed and range were increased to 65 mph and 250 miles, respectively. The disadvantages included refueling inconvenience, relatively short battery life (approximately 18 months), and barely adequate highway performance. Other advantages besides the obvious environmental qualities were exceptional fuel efficiency (approximately 47.5 miles per gallon for the combined diesel engine and fuel cell/electric motor) over the EPA urban driving cycle, ease of switching between systems, and mechanical maintenance and repair costs. The fuel cell was estimated to also have warranty and damageability advantages; however, these were not easily calculated since there were few production electric vehicles for comparison.

The design of experiments was created to measure the cost/benefit ratios of all of the above features and functions. The size of the fuel cell relative to the electric motor was one analysis. This trade-off also affected the cargo capacity and EPA interior volume, which were a measure of customer utility and potential loss of function.

Another design exercise on the UAV-100 involved the dimensional sizing and numerical variational control of the movable surfaces, including the rear pickup truck doors, rear lower sliding "pocket" panels, rear liftgate/tailgate and the upper cargo "pod." Computer simulation had modeled the fit-up tolerances and potential interference in production assembly. Certain characteristics of the material were suspected of potentially contributing to a "growth" phenomenon, whereby the exterior panels grow during high temperature cycles, resulting in interference for opening doors, and other parts. Material selection for the lower outer panels was modified to a composite having a compromised set of specifications to balance temperature stability as well as impact resistance, repairability, manufacturing cost, and mass.

The final consideration in the design phase of the UAV-100 was vehicle recyclability. The brainstorming session yielded the concept of a complete reconditioning, rebuilding, remanufacturing, or recovery of materials disposal facility, adjacent to the production plant. Studies were undertaken to analyze the economic benefits to salvaging vehicles for export to other markets or recycling mechanical or body subassemblies for repair of economically repairable vehicles, typically 8–12 years old and newer. The basis for the analysis was the depreciation tables, which showed that most vehicles were written off by their seventh year, unless their value was at least 50% above average. Efforts were underway to demonstrate that the new National vehicles were less depreciable than other competitive models and should be evaluated differently, based on the merits of the vehicle, resale value, and ultimate recyclability as a reconditioned vehicle with a full new car factory warranty.

The rest of the automotive industry scoffed at this concept, which had been tried before without success. A new approach was deemed necessary to convince the public that the new National program vehicles were in a class by themselves and needed to be evaluated on their value alone.

11.8.6 Manufacturing Design Input

The manufacturing engineering department had actively participated in the design team discussions and review sessions. They had identified several areas where improvements in the design could further contribute to increased efficiencies in the manufacturing phase, such as flexible tooling and fixtures, modular construction, powder- and water-based paint systems, and supply-chain integration. All of the above were proposals for cost savings sufficient to amortize the capital expenditures within the first one to two years of the production run. Suppliers who were encouraged to contribute to some of these new concepts were also rewarded with a portion of the profits and expense offsets. There were also a number of proposals to improve manufacturing and assembly operations. These included seat assemblies with side air bags, instrument panels, and door interior modules (commonly referred to as "plugs," since the assembly sequence required for these modules simulate simply "plugging" an opening). Dealer service concerns and warranty implications were being evaluated on these proposals.

11.9   PROTOTYPE DEVELOPMENT

11.9.1   C-147 Program

Power-train "mules" for the C-147 were easily assembled from production Centaur models. Rear floor pans were modified for the larger rear axle with integral torque sensor, cross member, jounce bumpers and increased ride height. The center floor pan had raised "bumps" for the rear driveshaft and exhaust system. New inner and outer body side aperture panels, a new roof panel, and the liftgate completed the major body components.

The engine compartment packaging was tighter due to the larger engine and transaxle, with the rear driveshaft hub and torque sensor. Both engine and transaxle electronic controllers were located in the left and right front corners, respectively, to minimize heat problems. The antilock brake controller was relocated under the master cylinder, due to the larger transaxle, to the left front corner, outboard of the rail. Unfortunately, this is where the majority of low-speed accidents occurred, based on damageability and repairability data from the insurance industry. While the nominal ride height was only a few inches higher than in the Centaur, bumper height was reduced to that of other passenger cars, as required by the Part 581 Bumper Standard. Larger headlamps and a larger grille opening, necessary for increased cooling, completed the front-end buildup. Bumper performance was designed to limit 5 mph damage to the bumper system and additionally, to the attached "crush cans," welded to the front rails, for the 9.4 mph 40% offset barrier low-speed insurance crash test. The front bumper beam was extended outboard 3 inches on each side to partially protect the controllers. In the rear, an additional reinforcement across the center section provided an extra margin of protection from center rear low-speed impacts. The rear rail crush cans were also designed to trigger or buckle at or above 10 mph from an offset impact. Most of the prototype components were fabricated in suppliers' model shops utilizing "hard" or production tooling, operated at slower speeds than those of regular production rates. Early prototype parts utilized one of the new stereolithography processes to produce nonfunctional but correctly sized and shaped underhood components to aid in packaging studies. Many of the new interior parts, such as trim moldings, were thermoformed from flat sheets of carbon fiber impregnated plastic. Seat back and cushion fabrics were manually machine stitched and stretch-formed over the prototype back and seat pans. Side air bags were fitted to the backs, along with the integral seat belts. New, stronger seat tracks were welded to the floor cross-car reinforcement on the center floor pan.

11.9.2   D-561/T-205 Program

Many similar actions were under simultaneous development in the D-561/T-205 program. Since these vehicles represented the company's first experience with separate frames, the supplier brought in several sample prototype frames for evaluation. Another supplier furnished body panels including floor pans, one-piece side apertures, and the pickup truck cargo box side panels. Front doors were common between the two models, as well as windshield and lock pillars. Four doors were being considered for the pickup, but could not be commonized with the sport utility because of the longer wheelbase necessitated by the separate cargo box, with the wheelhouse centered in it. The shorter-wheelbased sport utility had the rear door opening as part of the rear wheelhouse. Another T-205 derivative had an extended cab with quarter flipper windows and small folding jump seats. Two- and four-wheel drive, two cargo box lengths, and four bodies in the D-561/T-205 program provided the most comprehensive and challenging evaluation matrix of all the new National vehicles.

Fortunately, only one engine was offered, the supercharged 3.0-liter V-6, linked to either a five-speed manual or a four-speed automatic overdrive transmission. An automatic shifting high-low, part-time/full time transfer case and torque sensing differential, front, and rear completed the driveline buildup.

11.9.3   UAV-100 Program

The UAV-100 program, being all new in concept as well as design, presented the most complicated challenge for prototyping. Since the hybrid power-train development was delayed by lack of a workable fuel cell system, team members were assisting the C-147 group with their efforts to boost the 2.5-liter four-cylinder engine's output. Besides the variable valve timing mentioned earlier, some interest was being expressed in turbocharging. Two main problems were being explored through the prototype units: duty cycle and intercooling. Most turbocharged engines relied on boosting power at the midrange (i.e., 3000–5000) rpm. This was not, however, the ideal operating range envisioned for the UAV-100, where midrange started at 2000 rpm, and wide-open throttle was only 6500 rpm. If the exhaust-powered turbine was engaged at the low end of the range, it was deemed insufficient to boost power significantly. Also, extremely tight packaging and cost limited the use of an intercooler, but one was to be evaluated in case it permitted operation to be extended over a longer duty cycle. The possibility also existed that major mass reductions could eliminate the need for the turbocharger, but only if cargo capacity was also restricted, not a favorable trait in a utility vehicle.

The exterior body team was developing the plastic panels, utilizing several new material formulations, with the intent that at least one prototype vehicle would not only survive accelerated durability, but also perform well in the low-speed damageability and repairability tests. Several promised high-impact strength at minimum mass, but limited repair opportunities due to the incompatibility of the reinforcement fibers to most common adhesives. This meant that if minor scratches could not be repaired, the result would be panel replacement—and a major expense, which would have to be incurred by the insurance industry. Ultimately, if repair costs exceeded the range relative to that of average vehicles (i.e., greater than $\pm 20\%$), the vehicle would receive a surcharge for collision as well as comprehensive coverage since these were usually filed as a combined rating with the state insurance departments. This situation would render the vehicle noncompetitive for cost of ownership, another unfavorable factor in the purchase decision process. The material selection process was extended to find a better candidate.

An equally serious durability problem was envisioned for the variable wheelbase mechanism. Corrosion in the tracks caused the motor to bind up and stall, which could burn it out if not electrically protected. However, blowing fuses on a regular basis was not acceptable. Providing a moisture-proof cover for the tracks on the underbody, without adding significant mass and manufacturing complexity, was not seen as a feasible prospect. Another alternative design would have to be developed if the tests confirmed the suspicions of the team.

Crashworthiness was given top priority in each design program, with concern expressed over the new side-impact test and the performance of the air bag in the seat back, without any additional protection for the dummy's head. Several manufacturers were promoting "curtain" bags, which were deployed out of the front door header, in conjunction with the torso bag, from either the seat back or lock pillar. The side-impact test, utilizing a crabbed, moving barrier with a deformable faceplate, was supposed to contact the target vehicle in the middle of the front door. However, in several development tests with Orion models, the variance in the contact point was ±3.5 inches, resulting in unfavorable deployment characteristics (i.e., early or late timing) at the extreme limits. The location of the impact sensor was thought to be somewhat independent of the bag deployment timing, but this was not correct. The primary sensor was located inside the front door lock pillar, at the base, adjacent to the seat belt retractor. A secondary sensor was integrated into the system diagnostic module, located on the front floor pan, which also triggered the front air bags. Sensor development had progressed from pure mechanical switches, to hybrid rolling ball accelerometers, to elaborate three-dimensional strain gauges. Performance had also improved from several milliseconds to the microsecond level. The problem resolved into picking the most optimum location for consistent triggering.

The most inconsistent piece of the system was the response of the body structure, which depended on the force deflection properties of the sheet metal and plastics in the body. While the crush characteristics of thin-gauge steel is typically within ±10% of industry standards, spot weld quality was not consistently measured in production vehicles, due to the lack of a reliable nondestructive test procedure. Even magna flux and x-ray techniques, which usually provide predictable results for fatigue performance, have not always provided an acceptable level of reliability. This is due more to the nonlinear properties of large plastic deformations encountered in impact situations than to equipment accuracy, and makes predicting performance questionable and risky. Some success in modeling has come from the utilization of extremely sophisticated mathematical models on large, fast computers. However, the hardware tests have not come close enough to match the theoretical results. The team picked several sensors and alternate locations for crash testing.

11.9.4   Other Prototype Issues

Manufacturing got an early look at the new vehicles in the prototype buildup area. It was decided that union and management should work closely on the early models to uncover any problems that could still be solved prior to regular production. Most of the encounters involved small fit-up interferences that were solved by also having the suppliers on hand to correct them. Others were more serious in nature, requiring minor tooling modifications at the suppliers plant. Another outcome was a review of the complexity issue, in light of the mix of trim, color, and optional equipment combinations. The marketing department also had their representatives available to answer questions regarding the need for certain low-volume option loads that could potentially be eliminated without seriously disrupting marketing plans. For example, the upgraded optional tire on the C-147 was made the standard tire on the basis of the prototype evaluation, with no increase in price based on the volume guaranteed the supplier.

Cost estimates for mass and manufacturing complexity were calculated for comparison with earlier studies and forwarded to the new product committee for analysis. Several prototypes were loaned to the advertising agency for photography and copy layout work. Promotional themes were being created in addition to those already thought out from the market research clinics. After-sales parts analyses were also in motion, based on statistics from the parts department, on past part sales for current models for both mechanical/electrical and body items. The latter were confirmed by the insurance industry. Further feedback resulted in improvements in packaging large body panels, to prevent damage from handling and distribution at the warehouse. The marketing department developed retail prices for all base models and optional equipment packages and upgraded systems, based on final adjustments in the price quotes from suppliers. The final consideration in the design phase of the UAV-100 was vehicle recyclability. The brainstorming session came up with the concept of a complete reconditioning, rebuilding, remanufacturing, or recovery of materials disposal facility, adjacent to the production plant. Studies were undertaken to analyze the economic benefits to salvaging vehicles for export to other markets or recycling mechanical or body subassemblies for repair of economically repairable vehicles, typically 8–12 years old and newer. The basis for the analysis was the depreciation tables, which showed that most vehicles were written off by their seventh year, unless their value was at least 50% above average. Efforts were underway to demonstrate that the new National vehicles were less depreciable than other competitive models and should be evaluated differently, based on the merits of the vehicle, resale value, and ultimate recyclability as a reconditioned vehicle with a full new car factory warranty.

As mentioned earlier in the last chapter, the rest of the automotive industry scoffed at this concept, which had been tried before without success. A new approach was deemed necessary to convince the public that the new National program vehicles were in a class by themselves and needed to be evaluated on their value alone.

11.10   MANUFACTURING

The manufacturing engineering group had already voiced their concerns as part of the design review process. What remained included the proliferation of trim and color options on all of the programs. Plant end item complexity would be severely increased if all of the option packages were released as planned. Future plant quality audits predicted an overall reduction of 12–17% across all new model programs. Since warranty claims were currently running at 5% for the Centaur and 24% on average for the older models, plans to improve quality would not permit the new projections to be unchallenged. Several new product committee sessions were already scheduled in dedication of this issue.

A potentially far more serious quality issue was developing over the assembly and warranty problems associated with the UAV-100 variable wheelbase and movable side panel concept. The manufacturing engineering staff were attempting to model a quality function deployment scenario for the assembly sequence and had run into several areas of concern. The major exterior panel supplier being evaluated could not consistently produce parts within the specified tolerance for nominal dimensions of overall length, width, and thickness. These resulted from interference between the side panels and pockets where they were stored when the wheelbase was fully retracted. Also, when the hydraulic system started pumping more fluid into the actuators, the pressure surged from side to side, alternately binding and releasing, which caused a jerking motion and slightly damaged the rail tracks. Still another problem resulted in inadequate control of excessive whipping of the long rear brake flex hose, with the rear floor pan extension fully retracted, when the brakes were applied vigorously. This caused an excessive amount of brake pedal pulsation, which in turn caused uneven application of the brakes. This phenomenon was even experienced at the low speeds when simulating the movement of the vehicle in the plant, from the final assembly point to the marshaling area and during loading of the vehicles onto the rail cars and haul-away trailers. An alternative design was being evaluated in which the flex line could be routed inside the rear rails coiling out along with the rails, during the extension process. This in turn resulted in another concern, that of chafing the flex line cover material. Fatigue cycling tests were being contemplated; although the extension and retraction cycle was deemed to be a low frequency event, it was felt to be a potentially worse-case condition that may need to be checked.

11.11   PROMOTION AND DISTRIBUTION

Other issues that the new product committee and the four design teams considered as the final designs were being completed, were marketing plans, dealer relations regarding Internet sales, extended warranty coverage packages, and collision repair center/dealer development. Two issues from the dealer council were presented to the new product committee for resolution: Internet sales, and compensation for warranty claims from safety and emission equipment–related recalls.

National dealers were required to service any customer's vehicle regardless of the originating sale. In the last two years, nearly 15% of all National sales have been concluded over the Internet. The "dealers" involved were not real dealers at all but a storefront operation consisting of an office with a computer, modem, fax machine, and a telephones. Orders were received electronically over the Internet, after the customers completed their selection of standard models, optional equipment, exterior colors, and interior trim packages. Prices were displayed with a standard 15% markup to cover overhead costs and new vehicle preparation. Taxes, title transfer, and destination charges were extra. When the customer's credit application was approved, also electronically, and accepted by the customer, the order process was finalized, except for the standard three-working-day waiting period, during which the customer could change his/her mind and cancel the order without any cancellation fee. This streamlined ordering process resulted in savings at all points in the system. However, the dealers felt they were not getting their fair share of the profits from this new type of sales mechanism.

Based on the results of consumer research, it was determined that Internet sales would amount to 35–40% of new vehicle revenues and 20–25% of used vehicle revenues within five years. It was therefore decided that an Internet-specific marketing plan was necessary. Website developers, along with other electronic commerce specialists suppliers, were brought on board to assist in training and development of the plan. Virtual driving programs were envisioned to accompany 3-D animated showroom displays and enhanced sales information. Market beta testing was carried out and demonstrated several weak areas, which were clarified for final release. The entire program was scheduled for public introduction 30 days prior to introduction of the C-147 Centurion.

A problem resulted when warranty work was to be performed. The storefront dealer subcontracted new vehicle prep to independent shops since there was no repair shop behind the storefront. Customers who bought a car over the Internet had to take their vehicle to a regular full service dealer for warranty repairs. Naturally, the regular dealers resented this imposition since their own customers sometimes had to wait for routine service, especially when the recall concerned an important safety or emissions issue. Most regular dealers traditionally felt that full service was the best way to build long-term customer loyalty. They said, "If you took care of the customer in the back room, they would keep coming back to buy replacement vehicles, even during lean times." Several of the higher-volume regular dealers expressed the thought that these new customers were never going to come back to buy a new car from them, now that they could save money over the Internet and still get repairs and warranty work done at the local dealer. Reportedly, one particular dealer in a remote location had been selling several thousand units over the Internet, handling the prep in a local garage, and arranging for warranty service at one of 10 nearby dealerships by sharing some of the sale proceeds with the regular dealer. This was the type of creative sales management that the corporate office was contemplating across the country.

Dealers were not left out of the marketing plan, since they would still have to provide service and warranty repairs for all of the new program vehicles. Compensation was to be provided for the dealer located nearest to the buying customer. In addition, in rural areas, door-to-door service was to be available for minor repairs. For major repair work, dealers could subcontract it out to independent shops, if specific qualifications were met.

Collision repair could be arranged in a similar manner, assuming similar qualifications were in place for the collision shops. A major effort was established to have dealers get into the body shop business, including participation in the more important insurance company direct repair programs.

Extended warranty programs were also to be included and arranged to complement the standard dealer warranty packages. The main objective was to keep the dealer involved and responsible for doing warranty repairs accurately and timely. Full factory support was seen to be the ultimate insurance policy for continued and maximized customer satisfaction.

Emission control and safety equipment recalls were also to be handled differently, with a new streamlined expediting system designed to quickly verify the problem and a digital camera system to transmit images of the faulty parts to the warranty administration department, which would then authorize an electronic funds transfer to the dealer.

The other Internet-related problem was quotas. In the past, higher sales meant higher quotas of additional cars for the successful dealer. All dealers had to purchase a minimum quantity of new models based on their individual sales history and the economics of the particular region. If a new model started to sell well, dealers could get more cars by ordering more from the factory or by trading with other dealers who had surplus stock. If the plant capacity was limited, such as when the new model was outselling the competition, or when all models were selling in a boom economy, dealers frequently could not fill all orders and had to outbid each other for available units. This situation had occurred in the previous year, when the Centaur was introduced, with dealer bidding against dealer, resulting in customers being forced to pay 10–15% above the retail price. Now the situation changed: with trucks and other large, heavy vehicles selling well, the Centaur was no longer in as intense demand as it once was.

Large megadealers were forming alliances, buying other dealers, and selling "no-haggle price" deals every way possible, including over the Internet. They had earned the right to sell additional units based on their performance on past units. The regular dealers were not receiving additional units because they could not compete with the storefronts. Obviously, some additional compensation was needed for the regular dealer who performed the warranty work on the lost sale. The marketing department was attuned to the challenge and working on a solution.[3,4]

11.12   AFTER-SALES SUPPORT (INSURABILITY)

The service engineering team had found out that one of the benchmarked vehicles, Achilles, was noncompetitive from the standpoint of ownership cost, mainly due to insurance, the single largest portion of that cost. Two of the major components of the physical damage loss experience were collision and comprehensive, typically about 50% of the total insurance premium, the balance consisting of liability, medical, and property damage coverage. Collision damage, commonly called first-party coverage, re-

[3]Tim Moran, "Auto Industry Opens Its Private Internet," *Automotive News*, September 14, 1998, p. 22D.

[4]Frank S. Washington and Kathy Jackson, "GM Forming Its Own Network to Fight Online Buying Services," *Automotive News*, September 28, 1998, pp. 3, 63.

sulted from the vehicle impacting another vehicle, or a building or other property. Comprehensive consisted of losses from damage caused by or through other perils, including theft (the largest amount, approximately 40% on average), glass breakage (approximately 20% on average), fire, wind storm, hail, among other factors.

The Achilles model had nearly a 80% surcharge on its collision premium, and 150% on comprehensive. The collision loss experience was mainly due to the lack of repairability of the vehicle's separate frame, which the manufacturer deemed not repairable, even from minor damage, so as not to compromise the operation of the passive restraint systems, triggered by sensors mounted principally on the frame rails. The manufacturer had concluded that any heat associated with stress relief during straightening or welding would weaken the frame sufficiently to result in a loss of strength that could compromise subsequent operation characteristics of the air bag deployment.

The comprehensive losses were the result of the lack of any antitheft devices, either to delay illegal entry or drive away. Achilles were being stolen and stripped for spare parts or exported overseas to third world countries. The theft rate for Achilles was 4–5 times higher than for the average car, and 2–3 times higher than for the average sport utility. The manufacturer had recently released a highly regarded theft deterrence package, which consisted of an ignition immobilizer system. However, it was later learned that thieves were towing the car to a safe site for stripping, or for retrofitting it with an older Achilles or after-market ignition system. The vehicle was then driven or hauled to a port, loaded onto a ship inside a container, with phony paperwork showing it to be a shipment of manufactured goods, bags of cement or fertilizer.

The D-561 design team was assigned the task of coming up with a new theft deterrence plan, one not subject to the failings of the Achilles. Suppliers of new rolling code transponders were being evaluated for inclusion into key fobs for the optional keyless entry system. These systems prevented the vehicle from being started until the code transmitted by the transponder in the key fob was matched to the receiver in the ignition switch. The code was automatically changed every time the key fob button was activated. For base models without the keyless entry option, the antitheft system consisted of a mechanical device that, when tripped by attempting to remove the ignition switch with a slide hammer, would shut off power to the switch by means of a relay. Both of the above methodologies were to be modified to fit the other programs' ignition systems, so that all programs could benefit from the latest technology.

11.13   EPILOGUE

Chapter 16 contains an analysis of the various design parametric decisions and how they affected the final designs of each of the four team programs. The final outcomes were all compromises toward the targets provided for each design parameter. The general solution could have been applied to any automotive-type product, with the utilization of scale factors to reflect the level of design target in question. Actually, any product could be handled the same way, scaling the parameter ranges, as required.

One question that could be asked is, What is the most likely outcome of the four programs so far illustrated? Given that this treatise is somewhat theoretical, it could be surmised that the following might have happened.

11.13.1   C-147 Program

It could be argued that the small four-wheel-drive wagon was only a stop-gap measure until more marketable products were ready. Other manufacturers with limited resources have designed and built similar derivatives to existing products. Given the tooling investment constraint, there was little else that National could offer in the short-term. As it was, the Centurion would probably not have done very well, other than keeping National's name in the public view.

A more interesting issue compared to overall Centurion sales might have resulted from tracking the warranty performance of the drive train, when it was subjected to the higher stresses in all-wheel drive. Based on current utilization, four-wheel drive is not overly effective, mainly because most consumers have little knowledge of how to drive on wet, snow-covered, and slippery roads. Without a moderate dose of caution, many of today's drivers of large sport utility vehicles end up in some trouble, because they believe their vehicle can go anywhere at any speed. The fad of owning one of these behemoths passes quickly after getting stuck in deep snow or going into a ditch. Unfortunately, the market dictates what it wants, and the result is that large trucks are in demand.

11.13.2   D-561/T-205 Program

This program would have fared better than the C-147, but being smaller than existing market leaders, the niche for these vehicles would have also been small. Durability of the supercharged engine for maximum towing might have been problematic. The main and connecting rod bearings would probably have needed to be increased as part of an in-house quality improvement campaign, which might have been released after the first two years, assuming that warranty costs did not require it sooner.

11.13.3 UAV-100 Program

The most innovative of all of National's new products probably would have established a small but profitable niche, even without the hybrid power plant. However, the variable wheelbase concept would have been a failure, had the researchers delved into the surveys a little deeper. While many respondents liked the concept of increased utility, few would have bothered to shorten or lengthen their vehicles, because of the inconvenience of lifting and fitting panels, locking in the sides and top rear, or unloading and unloading seats. Most owners would have left the wheelbase either short or extended, without changing it, with the appropriate panels in place. Because the UAV-100 was small, load limits and towing might not have played a major part in any user activity. National could have done as well by designing two separate wheelbases with permanent roofs and side panels. The exercise accomplished one item, however; it tested several long-held but never verified theories concerning driver needs and wants for adaptability.

The above case study represents a complete examination of the new product development process, as viewed from the perspective of an outsider, contractor, investor, or observer. The view from an inside position would have been more difficult to comprehend and follow, due to the nature of being part of the organization and intimately involved in the daily work assignments. Hindsight is always clearer and cleaner than reality. The most important outcome to acquire from the case could be the discipline of the process itself, i.e., following the steps in a logical and interactive framework, accounting for the changes at each phase, and never losing sight of the overall product goals.

11.14 PRODUCT PLANNING PROPOSALS

The new product committee, together with the team leadership group, prepared a proposal for each new program. The Product Planning department developed the proposal format. A blank sample form appears in the appendix. Completed forms for each program appear at the end of this chapter.

11.14.1   C-147 Program

NATIONAL MOTORS CORPORATION

PRODUCT PLANNING PROPOSAL FORM: NEW VEHICLE PROGRAM/PLATFORM
PROGRAM/PLATFORM DESCRIPTION:
   PROGRAM HIGHLIGHTS:
   MAKE/MODEL/YEAR & JOB #1 TIMING:
      MAKE: NATIONAL   MODEL: CENTURION
      MODEL YEAR: 2001   JOB # 1 TIMING: 7/1/00
      PLATFORM DERIVATION: ALL NEW: MODIFIED C/O: X
         RESKIN: FRESHEN:
   STUDIO PROPERTY #/CHASSIS MOCKUP #:   C-147/D-7/C-31
   DRIVELINE FEATURES:
      POWER UNIT: 2.5 L I-4 W/NEW CYLINDER HEADS AND PLASTIC INTAKE
         MANIFOLD FOR INCREASED HORSEPOWER, (25-30HP DESIRED), AND
         LARGER EXHAUST SYSTEM W/REDUCED BACK PRESSURE
      TRANSFER UNIT(S):   FOUR-WHEEL DRIVE TRANSAXLE W/REAR EXIT PTO
         & SEPARATE SOLENOID FOR I/P ACTUATED PUSH-BUTTON SHIFTING
         IN/OUT OF FOUR-WHEEL DRIVE
      FINAL DRIVE UNIT(S):   TORQUE SENSING FLUID COUPLINGS AT ALL OUTPUT
         SHAFTS TO DIVERT TORQUE TO AXLE W/ MAX TRACTION. FRONT AXLES:
         HEAVY-DUTY CONSTANT VELOCITY JOINTS. REAR AXLE: H.D. HOTCHKISS
         JOINT W/ ONE-PIECE ALUMINUM REAR DRIVESHAFT.
   CHASSIS FEATURES:
      SEPARATE FRAME: CARRYOVER UNITIZED W/ ADDITIONAL STIFFENERS
         ACROSS FRONT AND REAR SHOCK TOWERS
      STEERING: POWER RACK AND PINION STANDARD
      SUSPENSION: CAST ALUMINUM UCA & LCA & COIL SPRINGS, FRONT & REAR
         W/ H.D. GAS SHOCK ABSORBERS STANDARD
      BRAKES: 4-CHANNEL ABS W/DISCS STANDARD
      TIRES & WHEELS: 6X15 ALUMINUM WHEELS & P175R15 ALL-TERRAIN TIRES
         STANDARD. FULL-SIZED SPARE TIRE STANDARD.
   BODY & TRIM FEATURES:
      BODY PANELS: NEW FRONT FENDERS W/ WHEELHOUSE BULGES TO
         ACCOMMODATE LIP FLARE MOLDINGS. NEW QUARTER PANELS W/ WHEEL
         LIP CONFIGURATION TO MATCH FRONT FENDERS & FLARES.
      NEW LOWER OUTER SILLS TO ACCOMMODATE MOLDINGS TO MATCH WHEEL
         LIP FLARE MOLDINGS, W/ SLIP JOINTS FRONT AND REAR FOR SEPARATE
         SERVICEABILITY & REPAIR/REPLACEMENT.
      NEW POWER EXTERIOR MIRRORS, NORMAL LEFT SIDE & WIDE-ANGLE (INSET),
         RIGHT SIDE. DAY/NIGHT INSIDE W/AUTO DIM FEATURE (OPTIONAL).
      PRIMARY STRUCTURE: UNITIZED STEEL, W/ CARRYOVER CENTAUR FRONT RAILS,
         DASH, FRONT FLOOR PAN, WINDSHIELD AND LOCK PILLARS, AND LOWER SILLS.
      SECONDARY STRUCTURE: NEW ROOF, INNER AND OUTER QUARTERS,
         LIFT GATE, AND CORNER POCKETS TO ACCOMMODATE NEW TAIL LAMPS.
         NEW REAR FLOOR PAN TO ACCOMMODATE SPARE TIRE AND LARGER FUEL
         TANK (MIN 20-GALLON CAPACITY).
      MOVABLE SUBASSEMBLIES: CARRYOVER FRONT & REAR DOOR ASSEMBLIES.
         NEW SMC HOOD W/ GRAPHIC DESIGN & STRIPING. NEW SMC LIFTGATE
         W/ GAS PROPS & WASH/WIPE PROVISION (OPTIONAL EQUIPMENT). NEW
         QUARTER OPTIONAL POWER FLIPPER VENTS. NEW MANUAL SUNROOF
         CENTERED OVER FRONT SEATS.

BUMPER SYSTEM PERFORMANCE & COMPONENTS: CARRY OVER CENTAUR
EXCEPT FOR OFFSET MOUNTING BRACKETS FRONT & REAR TO ACHIEVE
BUMPER HEIGHT REQUIREMENTS.

PAINT, COLOR, TRIM LEVELS & CONTENT:

BASE: CARRYOVER CENTAUR G SEATS W/ BLACK RUBBER FLOOR MATS
(OPTIONAL), OVER BASE CARPET & BLACK PLASTIC LOWER SILL COVERS.
VINYL SEAT COVERS, DOOR TRIM & KICK PANELS. BASE W/ CLEAR STANDARD
FOR ALL COLORS, INCL. BLACK, WHITE, SILVER, MED. RED, LT. BLUE & DK.
GREEN. SOLID COLORS ONLY (PPG TINT CHIPS ATTACHED). INTERIOR COLORS
TO MATCH EXTERIOR COLORS WITH TARTAN PLAID WEAVE FOR RED, BLUE,
AND GREEN. HERRINGBONE WEAVE FOR BLACK, WHITE, AND SILVER. CLOTH
OVER FIBREBOARD HEAD LINER & GLARE SHIELDS.

STANDARD LIGHT PACKAGE W/ MAP LIGHT OPTIONAL.

SECOND LEVEL: CARRYOVER CENTAUR GL LEVEL EXCEPT UPGRADED SEAT
FOAM ON FRONT SEATS ONLY. NEW CLOTH & VINYL SEAT COVERS AND TRIM
UTILIZING VOYAGER SERIES PRINT FOR ALL COLORS.
NEW COLORS INCLUDE CHARCOAL GRAY, MED BROWN, EGG SHELL WHITE,
LT. BLUE, DK. CHERRY & TEAL. EXTERIOR COLORS COORDINATED W/
INTERIOR COLORS.

THIRD LEVEL: CARRYOVER CENTAUR GLE-X, EXCEPT FOR ADDITIONAL SEAT
SIDE BOLSTERS AND FIXED AUTOMATIC HEAD RESTRAINTS.

NEW DELUXE INTERIOR FABRIC UTILIZING ADVENTURER SERIES, W/SOFT-WEAVE
PASTEL TRIM. PANELS COORDINATED TO SEAT MATERIAL.

REINFORCED SEAT BACK COVERS TO ACCOMMODATE MAP POCKETS.

FOURTH LEVEL: SIGNATURE SERIES W/ ALL NEW CONTOUR FRONT BUCKET
SEATS W/ NATURAL GRAIN COWHIDE LEATHER COVERS, TRIM PANELS,
CONSOLE AND HEADLINER. COLORS INCLUDE SUEDE BEIGE, ZEBRA PATTERN
(BLACK & WHITE), WHITE GOLD, CAMO PRINT (OLIVE & KHAKI), & CYAN
BLUE & MAHOGANY.

EXTERIOR COLORS INCLUDE PEWTER, CHAMPAIGN, TEAL, ICE BLUE, SAHARA
DUSK, & MEDEIRA RED. ALL COLORS ARE BASE/CLEAR/MICA TRI-COAT.

MATCHING LEATHER-WRAPPED STEERING WHEEL AND INSTRUMENT PANEL
APPOINTMENTS.

HEADS-UP DISPLAY ON WINDSHIELD BACK SIDE SHOWING SPEED, DIRECTION,
REAR & SIDE VIEW MIRROR (CCD CAMERAS), ADJACENT LANE OCCUPANCY,
SELECTED ENGINE PARAMETERS, ETC.

MARKET PERFORMANCE EXPECTATIONS:

MARKET SEGMENT: SMALL SPORT UTILITY

10% PENETRATION AFTER 1 YEAR

25% AFTER 3 YEARS

NEW ENTRY/REPLACEMENT OF: NEW ENTRY

MARKET-SHARE POTENTIAL: 3.5–4.0%

CANNIBALIZATION POTENTIAL: TBD

SUBCOMPACT CARS

COMPACT CARS

INTERMEDIATE CARS

FULL-SIZED CARS

MINIVANS

LARGE VANS

SMALL PICKUP TRUCKS

SMALL SPORT UTILITIES

MIDSIZE SPORT UTILITIES

LARGE SPORT UTILITIES
LARGE PICKUP TRUCKS
ANTICIPATED COMPETITIVE REACTION(S):
 BENCHMARK: NESTORER REDESIGN DUE 2002
 SECOND LEVEL: GR-3 REPLACED BY GR-4XL IN 2003
FINANCIAL PERFORMANCE EXPECTATIONS:
 VARIABLE PROFIT:
 BREAK-EVEN VOLUME:
 INCREMENTAL FIXED COSTS:
  HUMAN RESOURCE/STAFFING:
  PRODUCT DEVELOPMENT
  FACILITIES ALLOCATION:
 RETURN ON INCREMENTAL INVESTMENT:
DESIGN/ENGINEERING/MANUFACTURING/SERVICE EXPECTATIONS:
 DESIGN EXPECTATIONS:
 ENGINEERING EXPECTATIONS:
 PROCUREMENT EXPECTATIONS:
 MANUFACTURING EXPECTATIONS:
 SERVICE EXPECTATIONS:
 OTHER EXPECTATIONS: (i.e., DISTRIBUTION CHANNEL, ADVERTISING,
  INSURABILITY, WARRANTY, RESIDUAL VALUE AFTER LEASE, ETC.):

11.14.2   D-561 Program

NATIONAL MOTORS CORPORATION

PRODUCT PLANNING PROPOSAL FORM: NEW VEHICLE PROGRAM/PLATFORM
PROGRAM/PLATFORM DESCRIPTION:
 PROGRAM HIGHLIGHTS:
  MAKE/MODEL/YEAR & JOB #1 TIMING:
   MAKE: NATIONAL   MODEL: D-561
   MODEL YEAR: 2002   JOB #1 TIMING: 08/01/01
  PLATFORM DERIVATION: ALL NEW: X   MODIFIED C/O:
   RESKIN: FRESHEN:
  REF: STUDIO PROPERTY #/ CHASSIS MOCKUP #:   D-561/22.6/D-17.3
  DRIVELINE FEATURES:
   POWER UNIT: NEW 3.0 L V-6, ALUM. BLOCK W/ STEEL CYLINDER INSERTS.
    ALUM. HEADS W/ DOUBLE OVERHEAD CAMS, 4 VALVES PER CYLINDER &
    VARIABLE VALVE TIMING. SUPERCHARGING IS AN OPTION TO V.V.T.
   TRANSFER UNIT(S):   5 SPEED MANUAL GEARBOX STANDARD W/ 4 SPEED
    AUTOMATIC TRANSMISSION W/ LOCKUP TORQUE CONVERTER OPTIONAL.
    CHAIN DRIVE TRANSFER CASE W/ PUSH-BUTTON SHIFTING ANYTIME (EXCEPT
    LOW RANGE). TORQUE SENSOR FLUID COUPLINGS TRANSFER TORQUE TO
    DRIVESHAFT TURNING SLOWEST (I.E., MAXIMUM TRACTION).
   FINAL DRIVE UNIT(S):   CONSTANT VELOCITY JOINTS FRONT ONLY, HEAVY-
    DUTY HOTCHKISS IN REAR.
  CHASSIS FEATURES:
   SEPARATE FRAME: MILD STEEL, 100% BOXED
   STEERING: POWER RACK & PINION
   SUSPENSION: CAST STEEL UCA & LCA, FRONT & REAR LEAF SPRINGS. H.D. GAS
    SHOCK ABSORBERS FRONT AND REAR
   BRAKES: 4-WHEEL DISC W/4-CHANNEL ABS STD.
   TIRES & WHEELS: P255R75-15 RWL & 8X15 CHROME PLATED STEEL WHEELS

BODY & TRIM FEATURES:
BODY PANELS: STEEL @ MIN. GA. FOR MASS REDUCTION
PRIMARY STRUCTURE: FRAME & BODY STIFFENERS
SECONDARY STRUCTURE: ALL OTHER BODY PANELS, W/ HSLA REINFORCE-
MENTS, I.E., LOWER INNER SILL, INNER WINDSHIELD, LOCK & REAR QUARTER,
REAR CROSS MEMBER, INNER QUARTER & BUMPER BEAMS.
MOVABLE SUBASSEMBLIES: FULLY FRAMED, EXTENDING INTO ROOF HEADERS,
W/ DRIP RAIL MOLDINGS AND VENT WINGS, FRONT & REAR.
FOLDING, HEATED, EXTERIOR MIRRORS W/ ADJACENT LANE OCCUPANCY SIGNALS,
OPTIONAL. WIDE-ANGLE DISC ON RIGHT MIRROR.
COMBINATION LIFTGATE/TAILGATE CAN BE RAISED AND LOWERED AS ONE PIECE
OR TWO.
BUMPER SYSTEM PERFORMANCE & COMPONENTS: 5 MPH, PHASE I & TPO COVER,
HSLA REINFORCEMENT & CRUSHABLE MOUNTING BRACKETS
TRIM LEVELS & CONTENT:
BASE: VINYL-COVERED HIGH BACK BUCKETS IN FRONT, & REMOVABLE FOLDING
BENCH IN REAR. CLOTH HEADLINER ON FOAM BOARD. TRIM PANELS HAVE
VINYL MATCHING SEAT COVERS.
EXTERIOR COLORS ARE ALL METALLIC EXCEPT WHITE. OTHERS ARE MIDNIGHT
BLACK, MED. GRAY, BROWN, RED, BLUE, AND GREEN. INTERIOR COLORS ARE
BEIGE, CHARCOAL, MED. BLUE, CARMINE RED, AND DK. GREEN.
INSTRUMENT PANEL GAUGE PACKAGES INCLUDE BLACK ALPHANUMERICS ON
WHITE BACKGROUND, RECESSED VARIABLE LIGHTING, & ANTI-GLARE HOOD.
STANDARD GAUGES INCLUDE SPEEDOMETER, ODOMETER, FUEL, OIL
PRESSURE, COOLING TEMPERATURE & VOLTAGE.
SECOND LEVEL: CLOTH SEAT AND TRIM PANEL COVERS, TARTAN PATTERN,
HERRINGBONE, TWEED, & PLAID IN RED, GREEN, BLUE, & GOLD. BEIGE
HEADLINER AND REAR TRIM PANEL. CARPET UPGRADE TIGHT TWILL MINI-
SHAG IN BLACK LIGHT GRAY AND BEIGE.
WOODGRAIN VINYL INLAY ON INSTRUMENT PANEL AND CONSOLE, ARMREST
INSET, & DOOR HANDLE SURROUND.
THIRD LEVEL: LEATHER SEAT COVER AND DOOR TRIM PANELS IN SADDLE,
OFF-WHITE, CHARCOAL, & DARK BLUE.
MARKET PERFORMANCE EXPECTATIONS:
MARKET SEGMENT: MIDSIZED SPORT UTILITY
NEW ENTRY/REPLACEMENT OF: NEW ENTRY
MARKET SHARE POTENTIAL: 12% OF 7.25% MIDSIZED SEGMENT
CANNIBALIZATION POTENTIAL: MINIMAL
ANTICIPATED COMPETITIVE REACTION(S):   ACHILLES IS REPORTED TO BE
WORKING ON A LARGER DERIVATIVE.
FINANCIAL PERFORMANCE EXPECTATIONS:
VARIABLE PROFIT:
BREAKEVEN VOLUME:
INCREMENTAL FIXED COSTS:
HUMAN RESOURCE/STAFFING:
PRODUCT DEVELOPMENT:
FACILITIES ALLOCATION:
RETURN ON INCREMENTAL INVESTMENT
DESIGN/ENGINEERING/MANUFACTURING/SERVICE EXPECTATIONS:
DESIGN EXPECTATIONS:
ENGINEERING EXPECTATIONS:
MANUFACTURING EXPECTATIONS:

SERVICE EXPECTATIONS:
OTHER EXPECTATIONS: (i.e., DISTRIBUTION CHANNEL, ADVERTISING, INSURABILITY, WARRANTY, RESIDUAL VALUE AFTER LEASE, ETC.:

11.14.3   T-205 Program

NATIONAL MOTORS CORPORATION

PRODUCT PLANNING PROPOSAL FORM: NEW VEHICLE PROGRAM/PLATFORM
PROGRAM/PLATFORM DESCRIPTION:
  PROGRAM HIGHLIGHTS
    MAKE/MODEL/YEAR & JOB #1 TIMING:
      MAKE: NATIONAL   MODEL: T-205
      MODEL YEAR: 2002   JOB # 1 TIMING: 8/01/01
      PLATFORM DERIVATION: ALL NEW: MODIFIED C/O: D-561
      RESKIN: FRESHEN:
    DRIVELINE FEATURES:
      POWER UNIT: SAME AS D-561
      TRANSFER UNIT(S):   5-SPEED MANUAL & 4-SPEED AUTO
      FINAL DRIVE UNIT(S):   SAME AS D-561
      CHASSIS FEATURES:
      SEPARATE FRAME: 4X2 & 4X4 2-WHEEL BASES
      STEERING: RACK & PINION, HYD. POWER ASSISTED
      SUSPENSION: 4X2: FRONT COIL, REAR LEAF
                         4X4: FRONT TORSION BAR, REAR LEAF
      BRAKES: POWER DISC FRONT AND REAR, WITH 4-CHANNEL ABS STANDARD
      TIRES & WHEELS: SAME AS D-561
    BODY & TRIM FEATURES:
      BODY PANELS: STEEL EXCEPT COMPOSITE CARGO BOXES
      PRIMARY STRUCTURE: FULL FRAME WITH FRONT CRUSH AREAS IN FRAME
        HORNS
      SECONDARY STRUCTURE: HYDROFORMED STEEL FRONT UPPER RAILS,
        RADIATOR CORE SUPPORT W/ SEPARATELY SERVICEABLE UPPER TIE BAR.
      HYDROFORMED WINDSHIELD AND DOOR FRAME HEADERS AND PILLARS, IF
        POSSIBLE.
      MOVABLE SUBASSEMBLIES: FULL PERIMETER DOORS WITH WELDED IN SIDE
        RAILS AND ACCESS FOR INSTALLING WINDOW/DOOR HANDLE/ARM REST/
        TRIM PANEL PLUG. ALL MODULES SEPARATELY SERVICEABLE.
      CARGO BOX TAILGATE REMOVABLE BUMPER SYSTEM PERFORMANCE &
        COMPONENTS:
        PART 581 COMPLIANT, INCLUDING 4X4, WITH STANDARD REAR BUMPER
        AND CLASS 1 TRAILER HITCH SEPARATELY SERVICEABLE.
        TPO COVERED HIGH-DENSITY FOAM ABSORBER AND RIGID BRACKETS,
        BOLTED TO "CRUSH BOXES" WELDED TO FRAME HORNS
      TRIM LEVELS & CONTENT:
        BASE: SAME AS D-561
        SECOND LEVEL: SAME AS D-561
        THIRD LEVEL: AVAILABLE WITH SPECIAL ORDER
    MARKET PERFORMANCE EXPECTATIONS:
      MARKET SEGMENT:

NEW ENTRY/REPLACEMENT OF:
MARKET SHARE POTENTIAL:
CANNIBALIZATION POTENTIAL:
ANTICIPATED COMPETITIVE REACTION(S):
FINANCIAL PERFORMANCE EXPECTATIONS:
VARIABLE PROFIT:
BREAKEVEN VOLUME:
INCREMENTAL FIXED COSTS:
HUMAN RESOURCE/STAFFING:
PRODUCT DEVELOPMENT:
FACILITIES ALLOCATION:
RETURN ON INCREMENTAL INVESTMENT:
DESIGN/ENGINEERING/MANUFACTURING/SERVICE EXPECTATIONS:
DESIGN EXPECTATIONS:
ENGINEERING EXPECTATIONS:
PROCUREMENT EXPECTATIONS:
MANUFACTURING EXPECTATIONS:
SERVICE EXPECTATIONS:
OTHER EXPECTATIONS: (DISTRIBUTION CHANNEL, ADVERTISING, INSURABILITY,
WARRANTY, RESIDUAL VALUE AFTER LEASE, ETC.):

11.14.4   UAV-100 Program

NATIONAL MOTORS CORPORATION

PRODUCT PLANNING PROPOSAL FORM: NEW VEHICLE PROGRAM/PLATFORM
PROGRAM/PLATFORM DESCRIPTION:
PROGRAM HIGHLIGHTS:
MAKE/MODEL/YEAR & JOB #1 TIMING:
MAKE: NATIONAL   MODEL: UAV-100
MODEL YEAR: 2003   JOB #1 TIMING: 08/01/02
PLATFORM DERIVATION: ALL NEW: X   MODIFIED C/O:
RESKIN: FRESHEN:
DRIVELINE FEATURES:
POWER UNIT:
BASE: SAME AS C-147
OPTIONAL: HYBRID 1.5 L GAS W/ELECTRIC DRIVE (2004 MY IF FEASIBLE)
TRANSFER UNIT(S): SAME AS C-147
FINAL DRIVE UNIT(S): SAME AS C-147 EXCEPT REAR DRIVE ELECTRIC WITH
HYBRID POWER PLANT
CHASSIS FEATURES:UNITIZED BODY WITH VARIABLE WHEELBASE FEATURE,
HYDRAULIC MOTOR DRIVEN TELESCOPING REAR FLOOR AND RAILS.
SIDE PANELS STORED IN POCKETS BETWEEN OUTER AND INNER PANELS
BEHIND REAR DOORS. REMOVABLE REAR ROOF PANEL, STORABLE IN
REAR BED.
SEPARATE FRAME: NOT APPLICABLE
STEERING: RACK & PINION, POWER ASSISTED
SUSPENSION: FRONT & REAR COIL SPRINGS WITH UPPER AND LOWER
CONTROL ARMS AND ANTISWAY BARS
BRAKES: 4-WHEEL DISCS WITH 4-CHANNEL ABS
TIRES & WHEELS: SAME AS C-147

BODY & TRIM FEATURES:
  BODY PANELS: COMPOSITE, BOLTED ONTO HYDROFORMED HEADERS,
    PILLARS, RADIATOR CORE SUPPORT, LOWER SILLS
  PRIMARY STRUCTURE: CRUSH ZONES IN FRONT UPPER AND LOWER RAILS
    FABRICATED FROM DUAL-GAUGE TAILORED BLANKS. CONVENTIONAL
    SLIP JOINT SEAM TO BE LOCATED FORWARD OF SUSPENSION MOUNT.
    REAR SAME AS FRONT. PILLARS HYDROFORMED AND WELDED TO
    FLANGES TOP AND BOTTOM FOR ATTACHMENT OF OUTER PANELS.
  SECONDARY STRUCTURE: REINFORCED FRONT FLOOR PAN WITH CROSS-
    CAR BEAMS UNDER INSTRUMENT PANEL AND SEAT TRACKS
  MOVABLE SUBASSEMBLIES: HYDROFORMED DOOR RINGS, WELDED TO
    FLANGES SIMILAR TO SIDE STRUCTURE. WELDED SIDE BEAMS.
  BUMPER SYSTEM PERFORMANCE & COMPONENTS: PART 581 COMPLIANT
    WITH BEST IN CLASS RCAR LOW-SPEED OFFSET PERFORMANCE
  TRIM LEVELS & CONTENT:
    BASE: SAME AS C-147. IN ADDITION, CANVAS-COVERED TUBULAR FRAME
      SEATS, WITH MINIMUM FOAM PAD AND PLASTIC TRIM PANELS AND
      HEADLINER.
    SECOND LEVEL: SAME AS C-147
    THIRD LEVEL: NOT AVAILABLE
  MARKET PERFORMANCE EXPECTATIONS:
    MARKET SEGMENT:
    NEW ENTRY/REPLACEMENT OF:
    MARKET SHARE POTENTIAL:
    CANNIBALIZATION POTENTIAL:
    ANTICIPATED COMPETITIVE REACTION(S):
FINANCIAL PERFORMANCE EXPECTATIONS:
  VARIABLE PROFIT:
  BREAKEVEN VOLUME:
  INCREMENTAL FIXED COSTS:
    HUMAN RESOURCE/STAFFING:
    PRODUCT DEVELOPMENT:
    FACILITIES ALLOCATION:
  RETURN ON INCREMENTAL INVESTMENT:
DESIGN/ENGINEERING/MANUFACTURING/SERVICE EXPECTATIONS:
  DESIGN EXPECTATIONS:
  ENGINEERING EXPECTATIONS:
  MANUFACTURING EXPECTATIONS:
  SERVICE EXPECTATIONS:
  OTHER EXPECTATIONS: (DISTRIBUTION CHANNEL, ADVERTISING,
    INSURABILITY, WARRANTY, RESIDUAL VALUE AFTER LEASE, ETC.):

11.15   VEHICLE MASS COMPARISONS AND MANUFACTURING COST

11.15.1   Vehicle Benchmark Cost Distributions

Table 11.1 contains manufacturing cost distribution comparisons between major component groups for purchased and fabricated parts, fabrication and assembly labor, paint, and materials. The four new National Motors programs, as well as benchmarked competitive vechiles, have been included.

11.15.2   Vehicle Benchmark Mass Distributions

Table 11.2 contains several mass distribution comparisons between major component groups for the four National programs, as well as benchmarked competitive models, except for UAV-100. Other vehicle specifications are included, such as gross-axle weight ratings, GAWR, overall dimensions, and the number of seats and belts.

TABLE 11.1.

Vehicle description Unit cost category Purchased components	Baseline 4DR Sedan Centaur	Competitive Benchmark Vehicles				National Motors Design Programs			
		4WD Wagon Nestorer	Small P/U XP-50	Small S/U Achilles	Minivan Spanner	4WD Wagon C-147	Small S/U D-561	Small P/U T-205	UAV UAV-100
Parts Body	1,534	1777.06	552.07	2570.56	1259.16	1200	2000	500	1000
Chassis	716	834.8	376.45	962.37	734.2	600	750	300	500
Powertrain	2380	2502.34	1934.79	3065.86	1719.06	2000	2500	2000	1500
Mechanical	647	322.12	145.32	581.15	462.9	300	450	125	500
Electrical	311	476.29	86.54	235.49	194.83	300	150	100	250
Trim	444	589.53	289.4	622.89	649.06	300	400	300	250
Miscellaneous	104	93.9	52.56	42.5	63.81	100	50	50	150
Raw material Miscellaneous	75	74.75	27.51	67.8	31.98	85	50	75	25
Subtotal	6,211	6670.79	3464.64	8148.62	5115	4785	6350	3450	3950
Subassembly **Labor** Body	1388	1725.45	1944.87	1037.03	1288.15	1650	900	1800	1000
Chassis	103	257.9	379.65	418.36	119.84	200	350	300	500
Powertrain	353	609.87	598.05	895.4	722.04	550	800	500	1000
Mechanical	122	235.1	285.35	127.55	369.5	175	150	250	250
Electrical	81	128.23	56.13	256.78	157.9	100	200	50	250
Trim	56	103.5	82.34	314.91	87.65	100	300	100	250
Miscellaneous	27	35.07	12.86	52.5	46.09	25	50	50	250
Raw material **Subtotal**	432	567.19	826.57	193.14	651.45	550	700	500	250
Subtotal	2562	3662.31	4185.82	3295.67	3452.62	3350	3450	3550	3800
Paint Labor	175	207.88	236.45	268.17	192.67	175	250	250	50
Material	152	85.62	73.8	159.31	152.59	100	175	100	50
Subtotal	327	293.5	310.25	427.48	345.26	275	425	350	100

Final assembly									
Labor — Body	549	673.21	432.98	762.43	629.23	550	750	450	500
Chassis	550	622.7	768.24	809.55	690.41	550	850	800	1000
Powertrain	886	965.43	1119.93	1036.13	1229.85	875	1100	1250	1000
Mechanical	723	1009.34	253.4	954.92	256.88	900	850	200	500
Electrical	339	546.1	126	378.44	135.71	500	300	150	500
Trim	607	945.6	128.6	782.21	807.3	900	700	150	200
Miscell	38	50.58	16.08	46.37	29.65	50	50	50	100
Raw material	43	97.8	21.56	12.87	95.59	100	50	50	50
Subtotal	3735	4910.76	2866.79	4782.92	3874.62	4425	4650	3100	3850
Total direct labr	6624	8866.57	7362.86	8506.07	7672.5	8050	8525	7000	7750
Total direct cost	12,835	15537.36	10827.5	16654.6875	12787.5	12835	14875	10450	11700
Variable profit	2,246	2741.89	2985	3532.8125	3196.875	2246.125	2975	2873.75	3510
Dealer cost	15,081	18279.25	13812.5	20187.5	15984.375	15081.125	17850	13323.75	15210
Dealer markup	2,730	3225.75	2437.5	3562.5	3390.625	2865.41375	3123.75	2664.75	3802.5
MSRP base	17,811	21505	16250	23750	19375	17946.53875	20973.75	15988.5	19012.5
Tooling Costs — Body	12362479	15648017	10789642	17012540	12503465	15000000	20000000	10000000	3000000
Chassis	500603	1298605	3045167	766900	675108	750000	1000000	1000000	2000000
Powertrain	29675	57421	1829041	377895	524600	500000	400000	500000	500000
Mechanical	241256	367998	54623	89074	404050	400000	100000	100000	250000
Electrical	68879	72310	43890	117775	57600	75000	150000	100000	100000
Trim	734092	772311	125098	1175425	4794000	750000	1250000	150000	150000
Miscell	20985	33109	9834	53450	23670	30000	50000	50000	50000
Total	13957969	18249771	15897295	19593059	18982493	17505000	22950000	11900000	6050000
Fixed costs — Mfgplnt Facility allocat	2298565	9086943	24545000	38793000	28106950	13962400	20000000	30000000	10000000
Utilities allocat	423050	872130	284700	1786000	1016900	665000	2000000	3000000	1000000
Services allocat	488950	690562	378900	773450	485230	545900	1000000	1000000	500000
Equpmn depreciation	33152500	89743210	25104035	65725000	116547400	15675000	40000000	20000000	1000000

(continued)

TABLE 11.1. (Continued)

Vehicle description	Baseline	Competitive Benchmark Vehicles				National Motors Design Programs			
Unit cost category	4DR Sedan	4WD Wagon	Small P/U	Small S/U	Minivan	4WD Wagon	Small S/U	Small P/U	UAV
Purchased components	Centaur	Nestorer	XP-50	Achilles	Spanner	C-147	D-561	T-205	UAV-100
Mngmt salaries	23448755	10337450	17864000	19055000	19039800	6400000	12000000	12000000	8500000
Proper taxes	24895000	26980500	23110900	55000000	34145000	25000000	25000000	25000000	1500000
Miscellaneous	77980	104587	46385	210000	122800	125000	500000	500000	500000
Subtotal	84784800	137815382	94333920	181342450	62373300	100500000	91500000	36500000	
Engrg salaries	16187500	43462578	13784800	25675000	18615600	20234375	25000000	25000000	1500000
Testequ Deprec	5271025	11342687	3912050	8300000	6588800	7063173.5	10000000	10000000	5000000
Facility allocat	1350186	4504623	12654000	2460700	1485200	1660728.78	5000000	5000000	5000000
Utilities allocat	769893	1064310	591522	850000	885300	827634.975	1000000	1000000	1000000
Services allocat	834921	1690518	357430	605000	960200	897540.075	750000	750000	100000
Supplies	5109800	15473190	304100	3315000	5365300	5387250	5000000	5000000	250000
Miscellaneous	223450	577890	134500	125000	225000	250000	500000	500000	50000
Subtotal	29746775	78115796	31738402	41330700	34125400	37320702.33	47250000	42750000	2910000
Admin Salaries	8395000	16587900	4567000	14560000	8814750	8646850	10000000	10000000	5000000
Facility allocat	30178	77875	19340	350000	31700	31083.34	500000	500000	500000
Depreciation	115347	295670	14570	565000	126900	130342.11	750000	750000	7501000
Utilities allocat		84785	215628	55600	171500	89000	87328.55	200000	200000
Services allocat	134759	467824	8875	397000	141500	141496.95	500000	500000	500000
Supplies	65479	234785	2943	82000	75000	75300.85	100000	100000	100000
Miscellaneous	7449	7449	128600	749	706500	10000	125000	50000	50000
Subtotal	8832997	18008282	4669077	16832000	9288850	9237401.8	12100000	12100000	710000
Annual fixed costs	123364572	233939460	130741399	239505150	242878330	108931404.1	159850000	1.51E+08	7270000
Total fixed costs	493458288	1052727570	457594896.5	718515450	303597912.5	136164255.2	399625000	4.15E+08	2.18E+08
Total fixed tooling	507416257	1070977341	473492191.5	738108509	322580405.5	153669255.2	422575000	4.27E+08	2.24E+08
Volume capacity	321000	425675	250000	275000	1500000	125000	2000000	200000	100000
Breakeven	225907	390598	158623	208929	100904	68415	142042	148495	63860

TABLE 11.2.

Mass & specification study		Competitive Benchmark Vehicle					National Motors Design Studies				
		Baseline									
Vehicle type		2WDWGN	4WDWGN	SMALP/U	SMAL/S/U	Minivan	4WDWGN	SMALS/U	SMALP/U	UAV-100	UAV-100
Make & model		Centaur	Nestorer	XP-100	Achilles	Spanner	C-147	D-561	T-205	MIN WB	MAX WB
Mass in KG category											
Purchased components											
Front	Body	37.73	45.78	29.04	43.76	49.02	40	40	25	25	50
Rear		50.02	57.89	5.46	51.83	64.74	50	45	15	5	50
Front	Chassis	44.93	124.37	249.55	231.38	77.21	110	220	240	60	60
Rear		31.44	75.47	204.17	187.79	65.75	65	175	195	75	75
Front	Powertrain	15.28	29.3	11.45	18.37	41.6	20	15	10	10	10
Rear		10.89	13.16	4.9	5.49	32.47	10	15	5	10	10
Front	Mechanical	53.72	6052	59.87	61.94	58.23	50	50	50	50	50
Rear		18.87	23.41	26.02	22.31	34.93	20	20	20	20	20
Front	Electrical	9.07	14.74	7.29	12.9	15.07	10	10	10	10	10
Rear		4.51	10.18	2.91	13.71	7.89	5	10	5	10	10
Front	Trim & seats	43.7	52.77	13.98	36.94	56.75	45	30	15	15	25
Rear		58.27	71.88	3.86	27.62	49.35	60	25	5	5	25
Front	Fluids	28.11	30.92	31.96	33.17	29.63	25	30	25	25	25
Rear		13.85	14.07	15.38	12.05	16.49	10	10	10	15	15
Front	Miscellaneous	5.09	6.49	3.2	3.9	4.76	5	5	5	5	5
Rear		3.27	3.93	1.15	1.85	5.83	5	5	5	5	5
Front	Subtotal	237.63	358.4	406.34	442.36	332.27	305	400	380	200	235
Rear		131.12	269.99	263.85	322.65	277.45	225	305	260	145	210
Manufactured parts											
Front	Body	273.59	287.44	275.86	315.08	298.63	270	300	265	300	300
Rear		237.14	279.12	163.34	276.77	283.55	260	260	150	300	300
Front	Chassis	8.15	36.94	20.42	26.23	58.77	30	20	15	35	35
Rear		6.57	28.09	29.49	24.7	47.31	25	20	25	25	25

(continued)

185

TABLE 11.2. (Continued)

Location	Item										
Front	Powertrain	102.73	113.62	188.83	207.6	272.58	100	200	175	250	250
Rear		21.77	32.66	160.64	177.48	67.57	30	150	150	50	50
Front	Mechanical	11.24	15.23	12.07	8.96	14.65	10	20	15	20	20
Rear		2.36	8.11	5.62	7.23	10.82	5	15	10	15	15
Front	Electrical	3.14	4.59	3.34	2.57	10.42	5	15	10	15	15
Rear		0.79	6.45	2.14	1.6	11.23	5	15	10	10	10
Front	Trim & seats	1.46	2.67	1.83	12.57	37.56	5	15	5	15	15
Rear		4.75	6.51	0.79	15.09	55.21	5	15	5	15	15
Front	Miscellaneous	2.87	5.19	3.12	4.76	3.88	5	5	5	5	5
Rear		6.29	6.42	1.67	3.41	2.46	5	5	5	5	5
Front	Subtotal	403.18	465.68	505.47	577.77	696.49	425	575	490	640	640
Rear	Mfd parts	279.67	367.36	363.69	506.28	478.15	335	480	355	420	420
Front	Subtotal	640.81	824.08	911.81	1020.13	1028.76	730	975	870	840	875
Rear	Purch & Mfd	470.79	549.66	627.54	828.93	755.6	560	785	615	565	630
Total curb wt		1111.6	1373.74	1539.35	1849.06	1784.36	1290	1760	1485	1405	1505
Payload											
Front		122.50454	142.92196	166.9691	281.30672	245.0091	122.5045	281.3067	166.9691	281.3067	362.9764
Rear		183.75681	174.682396	607.9855	421.96007	326.6788	183.7568	421.9601	607.9855	544.4646	544.4646
Total	GAWR	306.26134	317.604356	774.9546	703.26679	571.6878	306.2613	703.2668	774.9546	825.7713	907.441
Front	GAWR	763.31454	967.00196	1078.779	1301.4367	1273.769	852.5045	1256.307	1036.969	1121.307	1237.976
Rear	GVWR	654.54681	724.342396	1235.525	1250.8901	1082.279	743.7568	1206.96	1222.985	1109.465	1174.465
Total	GVWR	1367.8613	1641.34436	2189.305	2302.3268	2231.048	1546.261	2213.267	2159.955	1980.771	2162.441
Total	GVWR in lbs	3014.7664	3617.52296	4825.227	5074.3282	4917.229	3407.96	4878.04	4760.54	4365.62	4766.02
Total curb	Weight in lbs	2449.9664	3027.72296	3392.727	4075.3282	3932.729	2843.16	3879.04	3272.94	3096.62	3317.02
Curb wt	Target in lbs	2450	3025	3395	4075	3925	2850	3875	3275	2975	3310
Specifications											
in mm	O/lngth	4495.8	4615.18	4927.6	4838.7	5041.9	4625	4825	4925	4225	4580
	W/base	2547.62	2710.18	2839.72	2783.84	2933.7	2700	2775	2840	2540	3300
	O/width	1549.4	1556.7	1752.6	1784.35	1739.9	1535	1750	1750	1750	1750
	O/height	1295.4	1379.22	1689.1	1790.7	1828.8	1375	1775	1685	1700	1700
Number	# Seats & belts	4	4	3	5	7	4	5	3	5	5

12

AERO CORPORATION

12.1 INTRODUCTION: A MILITARY AIRCRAFT MANUFACTURER MAKES THE TRANSITION TO THE COMMERCIAL MARKET WITH A REGIONAL/BUSINESS JET

This case study will follow a new aircraft model, throughout the 10-step new product development process, from strategic planning to after-sales support, utilizing Simultaneous Engineering principles in the most efficient manner possible. This is a phase-based process whereby engineers utilize advanced information technology to create a virtual design environment called the lean design process. The lean design process seeks to minimize recurring costs due to the mismatch between the quality of design engineering information and that of manufacturing. The charter given in the Product Life Cycle Process (Figure 12.1) states the Simultaneous Engineering philosophy applied to a lean design process format.

The Aero Corporation had been in the aircraft business for more than 50 years, producing large numbers of mostly military units for domestic as well as export markets. In later years, research and development efforts had resulted in significant subcontracts in space vehicle structures. Aero's expertise in power-plant development included both turbojet, turbo shaft, and

---

- Provide the Product Life Cycle Process model for products based on an understanding of customer needs and satisfying their expectations.
- This integrated process includes defining the product and means of production, validating and certifying the deliverable product, and supporting it in service.

FIGURE 12.1. Product Life-Cycle Process (PLC) Charter

bypass fan derivatives, through utilization of selected suppliers, who also designed and assembled jet and rocket engines for Aero's competitors. Basic manufacturing capabilities included extensive numerically controlled milling and boring machines for machining wings, fuselage, and control surfaces out of solid aluminum billets; and extensive semi-automatic fabrication equipment for smaller components, which needed to be welded and riveted together, such as engine nacelles, interior framework, equipment bays, access doors, and so on. Specialized vendors supplied avionics (communication, navigation, and weapons control instrumentation), landing gear, Plexiglas (plastic), canopies, tires and wheels, ground support (remote engine analysis condition and starting), and other modular (self-contained) systems.

Based on the current market situation, military aircraft sales projections showed demand for multipurpose fighters to be decreasing very quickly, even counting third world segments. Ground interdiction and support aircraft orders were already at a level below replacement, for accidental losses and fully depreciated write-offs. Orders for other military requirements were also decreasing rapidly. Aero Corp. had built a large number of medium-sized transports for cargo and paratroops over 20 years earlier, but the design had long since been obsolete. Remaining sales were to third world countries for military replacements, due to age, accidents, and attrition. The entire military infrastructure was contracting at a rate that would, very soon, result in a number of major plant closures and asset liquidations. Aero Corp. found itself in the midst of this quandary—of trying to decide what to do and how to do it.

12.2 STRATEGIC PLANNING

The strategic planning committee had met one week earlier and was now in the process of evaluating several possibilities for new ventures. One was to investigate a potential merger with another aerospace company, one with more of a commercial airframe specialty. Another was to diversify into another transportation business, while a third was to divest the firm and develop a new venture in an entirely new field.

The potential for a merger with another aerospace company was severely limited by available prospects. There had once been 50–60 large, well-capitalized entrepreneurial companies, some dating back 75–80 years—competently managed and profitable manufacturers and major suppliers of both military and commercial aircraft. Created in the cold war era of cost-plus-fixed-fee government contracts, the ability of the aerospace industry to pass along virtually any cost increase to taxpayers and customers, along with a constant demand by the airlines and the military establishment for new products, nearly full employment in the labor force, and generous support from elected officials involved in national defense appropriations, could not deter the inevitable decline and ultimate consolidation into the present five to six mega corporations, or multinational consortia. Tier 1- and Tier 2-level suppliers experiences a similar contraction, within certain specialties, namely large aluminum castings and forgings, instrumentation, landing gear, airframe subassemblies, and engines.

The airline companies also benefited from having pilots trained by the military, ground support logistics, personnel and equipment compatibility, and the development of designs that originated with military needs but were easily converted to civilian market applications.

Now, the market situation could be summed up as requiring desperate measures just to retain sufficient repair, maintenance, and refurbishing business to keep skeleton crews gainfully employed. Approximately half of all Aero Corp.'s production workers had been on extended leave for at least two years, and more were scheduled to go each succeeding month. Besides the aforementioned transports, several third world export contracts were the remnants of large all-encompassing orders for Aero's standard all-purpose fighter-bomber. Versions of it including reconnaisance, close ground support/interdiction, interceptor, long-range bomber escort, two-seat trainer, and electronic countermeasures were still being ordered, albeit in small numbers. However, these orders kept 30% of Aero's personnel busy on a full-time basis. Another 20% were engaged full-time in research and development projects, such as wind tunnel testing of new aerodynamic concepts, jet engine and rocket propulsion experiments, and new manufacturing techniques. These activities would soon have to also close down, since there was little revenue supporting them and few prospects for government research grants.

Several of Aero's direct competitors faced the same crucial decisions, in the next quarter. Two of them had commercial business, which was always more important than their military contracts. Multiple configuration standard and wide-body airliners made up most of the commercial business except for one unit that specialized in high-lift helicopters. Tilt rotor and other vertical or short takeoff and landing aircraft were designed specifically for geological research, exploration and oil well drilling and pipeline work. The other competitor specialized in high-capacity air freight conversions of former passenger airliners. Both companies were viewed as benchmarks for any new aviation venture undertaken by Aero, as well as potential partners for a new venture.

The early meetings had resulted in the committee voting to continue to utilize Aero's capabilities to develop aircraft, particularly commercial types, based on the expertise—either in-house or available outside from suppliers or potential competitor partners—and the assets available, without a serious recapitalization. The chief finance officer's staff had surveyed their major investors to test the above decision and found that most were favorable to it. There was some concern expressed over the firm's lack of expertise in the commercial market, but the investors were willing to let the firm find its way into the innovation phase, without any interference. The board of directors had major investor representation and did not hesitate to offer advice and technical expertise if required or deemed necessary. They reserved the right to challenge any new product/marketing plans which could not be adequately justified for either new, profitable segments of existing products, or all-new product ventures.

12.3 MARKET RESEARCH

The marketing department had been conducting research clinics with major airline executives and longtime business customers, to find out their needs and wants for the next decade. It appeared as though a major market opportunity existed in the regional jet segment. This translated into several smaller subsegments of different-sized aircraft, designed specifically for the operational envelopes listed below:

# Pass./LF[a]	Max. Range (NM)[d]	Cr.[e] Speed (NM/hr)[f]	Max. T-O/LNG WT (lb)[b]	Fld. Length[c]
100/50	1000	425	55,000	5000/3500
50/35	1000	425	45,000	5000

[a]Breakeven load factor
[b]Takeoff and landing weight
[c]Takeoff and landing field length
[d]NM: nautical miles
[e]Cr: cruising speed
[f]NM/hr: nautical miles per hour

The other opportunity area existed in the convertible cargo/business jet segment, where the same aircraft could be quickly converted into a combined passenger and cargo configuration or an all-cargo version. Customized interiors could further convert the same airframe into a business/executive/charter jet for overnight travel, extended trips, or vacations. The challenge was to find out how much convertibility could be designed in up front and its relative merits, demand and cost.

The risk assessment for the worst-case scenario had been calculated, and it showed only a slight chance of complete failure (i.e., sales below breakeven), and a moderate chance of selling several hundred units of several types. However, the analysis was based on preliminary market forecasts and current orders for similar aircraft. Demand for the type of aircraft envisioned for new regional carriers, worldwide, could amount to several thousand units, over the life-cycle constraints utilized and the assumptions of market growth and economic stability. Obviously, other airframe manufacturers also were targeting this same market segment for their products and were thought to be developing very aggressive plans to exploit this opportunity.

Major airline executives were informally contacted by board members to ascertain their current operational concerns, problems, and future requirements for regional and feeder line equipment. A general picture of the market situation was beginning to come into focus, with several potential segments underdeveloped and underutilized by the public, small package shippers, and business executives. What was needed next was an comprehensive market research study to fully develop the concepts and expand what was initially a new product proposal to the board into a presentation, scheduled for the next quarter, which would cover virtually all aspects of a new regional/convertible transport.

The research department then contracted with a service provider to develop a set of surveys to address the concerns and collect the ideas from all affected groups, including customers and employees. The survey questions covered virtually all topics affected by and affecting air travel, in areas such as passenger safety and airframe crashworthiness, operational characteristics, comfort, convenience, performance, baggage handling and retrieval, food and beverages, ticketing, airport congestion, parking, scheduling noise, refueling, maintenance, and repair.

12.4   IDEA GENERATION AND SCREENING/EVALUATION

The analysis that was produced from the surveys provided an opportunity to create a new product to fit one or more of the niches that were revealed by the data. The major complaint echoed by customers was the aggravation stemming from the inconvenience of the airline schedules, along with the required accommodation that the schedules demanded. Ground traffic congestion of not only vehicles but also aircraft was another concern, as was baggage. Most passengers liked to be able to carry on all of their baggage, but realized that this choice was increasingly becoming unattainable due to storage and safety reasons. The thought of endlessly taxiing around the airport or waiting for a gate was also criticized as unnecessary. The practice of aircraft holding in a pattern while waiting for landing clearance was not perceived as a big problem, since most occurrences were thought by customers to be weather- or traffic-related. Crash survivability, however, did emerge as a major concern, never before seen in large numbers. All participants expressed some interest in recent airliner crash investigations and the possible outcomes on future aircraft designs for improved safety system redundancy and crash survivability. In-flight engine noise ranked approximately the same as jet takeoff and landing noise in the vicinity of the airports. Overall, most customers were concerned about the impersonality and mind-numbing reactions of having to deal with large numbers of other travelers, baggage, and check-in lines in the big hub airports. Many respondents wondered why there were no other alternatives to the hub and spoke flight management systems.

12.5   PRODUCT PLANNING

Based on the survey analysis and their own internal studies, the product planning group, a subcommittee of the new product development committee, began the process of developing a proposal, for a new regional/convertible jet program, based on results of the idea generation and screening processes. Since it was unclear which power-plant configuration would be chosen (i.e., prop jet or fan jet), both types were analyzed, and preliminary plans for them developed.

The key factors are the ability to takeoff and land on relatively short runways (5000 ft max) from smaller local/regional airports, with equipment designed to carry 30–100 passengers (two separate engines, jet and prop), along with a variety of convertible cargo or business jet configurations. Maximum range would need to be approximately 1000 nautical miles (nmi), but the average flight leg would probably be 250 ± 100 NM. This would reduce turnaround time for refueling to every two to three legs instead of after every flight.

The main reason for considering equivalent capacity aircraft with different power plants was the anticipated opposition to large increases in takeoff and landing jet noise at local and regional airports. Some of the smaller airports, in large metropolitan areas, had been in existence for more than 50 years, subsiding generally on the fees charged for handling private or business airplanes, with the actual number of takeoffs and landings in the same range as those at major airports.

Due to the proximity of the airports to residential areas, which in most cases had reached maturity long after the airport had been established, there were forces automatically opposed to any airport expansion. The Federal Aviation Administration (FAA) controlled the air space within the United States, including the vicinity of airports. This meant that if approved, carriers could operate scheduled flights from small airports.

Political pressure and lobbying from homeowner associations, realtors, school boards, churches, and other groups had convinced most small airport management that commercial expansion, utilizing jet engine–type aircraft, would be vigorously opposed, resulting in an almost impossible mission to convince the public, either as bystanders or potential customers, that expansion of airport operations was good for everyone. While there were sufficient numbers of business flyers to warrant development of alternative air transport modes, there may not have been sufficient numbers of politically influential flyers to challenge the current status.

Fortunately, sufficient demand existed in many rural areas welcoming new, alternative air travel to negate the reactions of suburban voters. Having two to three regularly scheduled flights each day, each way, from a rural regional airport to a major hub offered a better option than spending 3–6 hours commuting by bus to the nearest hub facility, only to stand in long lines waiting for check-in or boarding passes. The regional aircraft concept offered travelers a short, comfortable flight and automatic ticketing to their main connection, baggage transfer routing, less stress, and so forth. Aero senior management felt very positive about this concept and was eager to pursue it.

During the previous brainstorming session, several conceptual design alternatives were proposed, requiring further development:

1. Swept-wing, two-jet engines, placed on rear pods, attached to a typically configured long, slender fuselage, and high vertical fin and T-tail.

2. Straight, low-wing twin turboprop, with engines mounted in the mid-wing location and conventional horizontal and vertical tail.

3. High straight-wing configuration with turbojet engines hung on nacelles below with high T-tail above fuselage.

4. Same as #3 with turboprop engines.

5. Same as #3 with short takeoff and landing (STOL) capabilities (i.e., triple-slotted rear flaps, leading edge slats, limited vectored thrust).

6. Same as #4 with STOL performance, except propellers and engines are geared together for more favorable one-engine-out capability.

7. Each alternative would have three fuselage lengths, for 10–20, 50, and 100-passenger capacity and convertible for 25% and 50% cargo capacity (passenger versions only).

8. Each alternative would have up to five optional business executive interior packages for each fuselage length.

Each alternative was assigned to a team around a separate airframe configuration, with engine differences grouped together within the same team. Numbers 1 and 2 were separate entities, design teams 1 and 2, respectively. Numbers 3 and 4 were grouped together as design team 3, while the STOL efforts, 5 and 6, were grouped together as design team 4. All derivative modular fuselage length, cargo conversion, and interior package developments were incorporated into the same teams as the base airframes. The benchmark for team 3 was an existing airframe originally designed for turboprops and later adapted to jet engines. Teams 7 and 8 were to operate as evaluators for determining whether the same fuselage modules could be applied to all three passenger cabin lengths.

Team members included advanced manufacturing experts, charged with developing the most cost-effective construction methods. This included investigations of new materials, fixturing, joining, and assembly techniques. Purchasing representatives were scanning their suppliers' contacts for new opportunities and specialized talents in adhesive technology, glass fiber impregnation, and aluminum alloys.

Adaptability or upgradability was viewed as a potential design issue. While turboprop-equipped regional aircraft have traditionally had better direct operating costs (DOC) than jets, the public perception is that of an old-fashioned, slow conveyance compared with jet-powered aircraft. This attitude is also based on the typical "puddle-jumper" image of propeller planes, with small, cramped seating and noisy, nonpressurized operation. Jets were and are considered more "modern, sleek, fast, and efficient."

The cargo convertibility concept was created to fill a void in the more remote or rural markets, where small package delivery operations, medical express, and even priority mail services were interested in the low shipping rates that could conceivably result from regularly scheduled passenger runs. One of the key elements identified in the market research was concerns over cargo movement during rough-weather flights, and increased noise levels or odors emanating from adjacent, unsealed, uninsulated, or underinsulated cargo compartments. Specifications developed for the three passenger configurations—70, 50, and 20—are shown in Table 12.1 at the end of the chapter.

12.6   DESIGN AND ENGINEERING

From a design perspective, concern was expressed about a foolproof movable wall system that could be easily reconfigured, even as passengers wait to board the next flight, with the seat pitch and wall changes occurring in their full view. Later it was resolved into a scenario in which all configuration changes would necessarily have to be performed in maintenance hangers.

Short takeoff and landing (STOL) capabilities had been developed for military applications over two decades before, but the results were not promising. Excessive weight, tedious operational characteristics, (i.e., noise, vibration, and harshness, or NVH), manufacturing complexity, and service penalties were encountered. The military later progressed to full vertical/short takeoff and landing (V/STOL) systems as exemplified by the McDonnell-Douglas AV-8D Harrier vectored-thrust ground support fighter and the new Boeing V-22 Osprey/Bell 609 military/civil tilt-rotor, further abandoning conventional STOL ideas as impractical. However, in spite of these restrictions, the STOL alternatives were left in the design study, if for no other reason than to challenge the team's creativity. Because of the need to develop the ideas that were generated, the screening/evaluation process step was postponed until the concept designs could be completed.

The benchmark aircraft selected were as follows:

Aircraft Description	Configuration (Wing, Engine)	Mission Category
Fokker/Dasa 70/100[a]	Low-swept rear pod	70/100 passenger jet
DHC Dash 8 series 400[b]	High straight below wing	100 passenger prop
Canadair Regional Jet[c]	Low-swept rear pod	50 passenger jet
Fairchild Dornier 328[d]	High straight below wing	50 passenger prop
Beech 1900D[e]	Low straight midwing	10–20 passenger bus, prop
Dassault Falcon 2000[f]	Low-swept rear pod	10–20 passenger bus. jet

[a] Jackson, Paul, et al., *Janes's All The World's Aircraft*, Jane's Information Group, Alexandria, Va., 1996. pp. 40–41.
[b] Ibid., pp. 106–107
[c] Ibid., pp. 29–31
[d] Ibid., pp. 294–297
[e] Ibid., pp. 709–710
[f] Ibid., pp. 92

The concept designs were then begun, with the teams dividing up the initial sizing work tasks and assigning specific items to experts in aerodynamics, propulsion, weights, flight mechanics, avionics, landing gear, ground support, and so on. Preliminary aerodynamics and performance and weight estimates were also part of the concept development phase. Some of the more important work involved calculating empty and maximum takeoff weights, thrust-specific fuel consumption, thrust/weight and wing loading ratios, lift/drag ratio, ceiling, lift and drag coefficients, and other parameters needed for estimating performance, stability, and control. Many iterations were typically required to progress from initial to final design configuration and shape. The teams agreed to share certain modular aspects of the overall design to save time. A summary of intended design actions follows.

High bypass ratio fan jets from the same family would power all jet versions, varying in thrust and fuel consumption, as required by the initial sizing calculations. A common family of turboprop combinations was being considered for the prop alternatives, although the business plane would probably meet its flight envelope objectives with smaller engines than would be needed for the larger, heavier-weight passenger planes.

Three standard fuselage modules, of constant cross section, were being designed. The first module was forward of the wing, with wardrobe closet, toilet, and galley. The front module was mated to the standard cockpit module using a common flange arrangement. The second module was to be integrated into the wing box, attaching to the wing spars, in a socket/sleeve mount, designed to be interchangeable for either high-wing or low-wing mount configurations. The third module, aft of the wing, included large cargo doors on either or both sides, rear galley, toilets, and the empannage or rear fuselage taper and vertical tail interface.

There were also standardized systems specified for pilot controls, which were designated as direct cable with hydraulic assist, avionics (including Heads-up Display [HUD][1]), fuel systems, and flight deck. Galley, toilets, seats, luggage, and baggage areas were designated as "standard" modular units, to allow interchangeability between the 50- and 100-passenger and cargo conversion configurations.

The business jet interior modules varied by level of comfort, from basic weave fabric–covered upholstery and plastic fittings to leather, wood, and glass ornamentation. Executive suite furniture was also available is several shades and textures with wood laminates, aluminum framework, and

---

[1] Altitude, speed, and flight angle display on backside of windscreen.

high-quality porcelain bathroom fixtures. Convertible couches, with sleep restraints, completed the furnishings.

The basic airframe called for a two-spar wing, with flush skin, internally reinforced with bonded stringers, stamped and riveted reinforcements for fuselage panels, access panels, surrounds for windows, and so forth. An alternative investigation was initiated into one-piece hydroformed fuselage "tube" sections and a one piece stamped "ring" for engine nacelles and vertical and horizontal tail structures. Major fuselage sections were to be weld bonded, along with welded reinforcements, between the cockpit and the front passenger module, and between the center, rear, and tail sections. Plug-type passenger and service doors were specified for the left and right front area, aft of the cockpit, with large, cargo access doors in the rear, side forward of the engines. These large doors were for facilitating the conversion to combined passenger-cargo interior modules, or for unloading and loading cargo on special pallets. All necessary electrical, hydraulic, plumbing, and communications lines were to be installed in all aircraft, with standard connectors between modules for future utilization in alternative modes.

The "standard" flight envelope specified for the 100-passenger version (hereafter referred to as the Dash 100) begins with a maximum takeoff weight of 45,000 lb, maximum takeoff field length of 5000 ft, a maximum rate of climb of 3500 ft/min, a normal cruising altitude of 35,000 ft, and a 1000-mile range. Field length for maximum landing weight was to be 4500 ft. Leading and trailing edge flaps were to be deployed for every takeoff and landing, along with combination spoiler/lift "dumpers" for decreasing lift and increasing drag after touchdown. Winglets were to be investigated if additional winglift was required. Ailerons on the wings and elevons on the horizontal tail were specified for maximum control flexibility. A vertical tail with combination rudder/"spoileron" was thought to be an advantage for high-angle-of-attack maneuverability during cross-wind landings, or after takeoff, particularly from small intercity airports.

The Dash 50 configuration (50 passengers maximum) had similar specifications, with lower maximum takeoff and landing weights, proportionately shorter takeoff and landing field lengths, and a slightly lower range, although increased-range fuel tanks could be added without incurring a severe performance penalty.

In order to comply with local noise ordinances and FAA directives, operational levels were specified at 75EPN (effective perceived noise) decibels (dB) for takeoff, 90 dB for approach, and 80 dB for sideline measurement. It was assumed that a major campaign would be necessary to overcome the concerns of local citizen groups on not only noise but also jet fuel odors, emanating from runways and taxiways into residential neighborhoods adjacent to regional airports.

The main theme for the utilization of these fields was to avoid the congestion, hassle, traffic, and stress of the larger hubs. Parking lots were envisioned to be within easy walking distance to the ticket areas, with luggage vans at the parking lot and curbside check-in, luggage handling, and automatic ticketing available. Normal ticketing, check-in, and luggage handling at the ticket counters would also be available. Gates would also be relatively close (i.e., no more than a 10-minute walk unpowered), with automatic check-in at the gate for passengers without luggage. Carry-on luggage would be limited to two pieces, of a maximum size and weight.

Departure and run-up times were to be minimized by the proximity of the taxiways and runways. Takeoff, climb out to reach regional control, and time to reach cruising altitude all offered advantages for a smaller regional aircraft. Four-abreast seating with a center aisle offered the advantages of generous spacing and no middle seats. Maximum passenger loads on minimum distances would require two flight attendants, working from each end.

Small galleys were to be located in either forward or rear passenger modules, on the starboard (right) side, with service doors to facilitate contract vendor servicing. Standard galley equipment included microwave ovens for reheating cooked meals from the vendors. Optional galleys could be ordered to replace the standard galleys with convection ovens, refrigerators, and cooking tops for regular meal preparation.

The business jet configurations included five different sets of interior decor, to be developed in conjunction with a business jet customizing supplier, in consultation with a panel of primary users (i.e., chief executives from several different industries). Generally, leather-covered seats with swivel bases, tilt backs, and thigh and leg support were provided, along with a storable conference table or work surface, lighting, and a seat-back surround-sound system. Sideboards and cabinetry were to contain safe stowage for glass and tableware, linen, and silverware. Beverages were to be available from refrigerated compartments, also located in the office areas. Optional deluxe galleys were to be designed for gourmet meal preparation. Twin-size beds were to be designed as convertible from the three-passenger lounge seats, complete with restraints for rough weather flying. Accommodations ranged from 4–10 sleeping berths, or 12–18 seating positions, depending on the other equipment specified (i.e., computer work stations, secretarial or other office facilities). Special requirements were seen to be necessary to counteract excessive electromagnetic interference from onboard computer and other office equipment, located in the passenger compartment. A new type of shielding was to be investigated.

Cargo conversion was to involve removal of standard seat modules, through the cargo doors, leaving the seat tracks. Special pallets were to be designed to roll on and off without damaging interior module or seat track mountings or interior walls or ceiling. Loaded pallets were to be automatically weighed prior to loading and the total weight tallied in progress. Balance information was to be also automatically calculated on the ground prior to loading so that the proper sequence of pallets was loaded (i.e., lightest rearward, heaviest forward). In-flight tie-downs were to be standardized for location and strength, so that any pallet, regardless of load type or distribution, would not be affected by the tie-down location.

Passenger, baggage, and cargo handling facilities at small local and regional airports could end up affecting airframe or infrastructure decisions. A sudden increase in passenger traffic could severely tax ground facilities such as fuel supply, security systems, baggage handling, toilet capacities, and parking. Sometimes, these airports are administered by local city, county, or combined governments, which might need voter approval prior to issuing revenue bonds for improvements geared toward new airline services.

With the advent of airline deregulation, economic decisions, as well as the FAA, governed where airlines chose to fly. While there would no doubt be some economic benefits, increases in traffic, noise, and air pollution might not justify the added ticket sales-related revenue in the eyes of the majority of voters, who typically do not fly, except on rare occasions and then usually to vacation resorts, beyond the range and immediate interest of regional airliners and airline companies. The potential for some feeder line traffic into the larger airports and hubs was not overlooked but seemed to represent a minor consideration during this phase of the analysis.

The above represented the results and analyses of market research of a different sort, but it is as important as the results of customer, airline employee, and aircraft worker surveys. The pulse and attitudes of nearby residents and of government and political leaders needed to be assessed, both with and without consideration for the political makeup and timing of local, regional, and state elections.

Despite certain reservations expressed in the surveys regarding turbo-prop airplanes, it was decided that at least one would be included in the design study. There had been several customer inquiries and concerns over noise as well as the safety of turbo-propeller aircraft, particularly when passengers are sitting in window seats adjacent to large spinning propellers outside their seats.

Consideration was also encouraged for other infrastructure issues affecting potential future profitability of regional air routes. Fuel availability, storage, delivery and cost, takeoff and landing fees, relocation of airline personnel, equipment maintenance, personnel and storage facilities, FAA radar-controlled field equipment and personnel availability, and interference from takeoff and landing approaches from larger airports—all needed to be evaluated while compiling a business case for Aero senior management final approval and the development of a presentation package to the airlines and local airport supervisory boards.

The initial sizing of the alternative concepts included the following calculations:

Takeoff weight buildup
Empty-weight estimation
Fuel-fraction estimation
Mission segment/profile weight fractions
Specific fuel consumption
Lift/drag estimation at wetted aspect ratio
Maximum total weight calculation

After analysis of the initial sizing calculations, it was determined that a modular design airplane could be built to meet all of the mission requirements, with reasonable fuel consumption and an economical flying envelope and range. The next phase was for the detailed design work to be started. It included all of the iterations required to generate a final design, performed more or less simultaneously. (See Table 12.1 at the end of the chapter for some of the more important Aero aircraft specifications.)

12.6.1   Trade-off Studies

As the conceptual design studies progressed, a number of "trade-off" studies analyzed some of the following decision points besides initial aircraft sizing calculations:

Jet vs. prop power plants
Passenger/cargo feasibility, conversion time, and location
Modular fuselage construction
Material selection criteria
Life-cycle costs

The analysis of power-plant selection was as much an emotional issue as a technical one. Turboprops had clearly superior fuel efficiency, but lacked the cruising speed and smoothness of jets. Customers, in recent market research clinics, preferred jets by a wide margin, nearly 3 to 1. These was genuine fear about propellers breaking and spinning through the fuselage, even though propeller failures were extremely rare. Then there was the debate on jet noise, which occurred in front of the board of supervisors of a nearby regional airport. Aero executives, along with representatives from a new feeder airline, Omni-Central, made a presentation on the Aero Dash series, in order to address the concerns of people who live near the airport about not only noise, but also the increase in takeoffs and landings. The presentation included concept drawings of the new aircraft in the 50-passenger version, a layout of the new terminal building, including ground service facilities, and enlarged and upgraded parking lots with car-to-gate shuttle minivan service. While $265 million in municipal bonds would be required to pay for the cost of new taxiways and tarmac for the loading and unloading of passengers and baggage, the impact on property taxes was shown to be minimal, due to the taxes on airline revenue and fees to be paid for every landing. A videotape was then shown featuring several competitive regional jets operating from other airports. It showed planes, such as those in the benchmark study, taking off, landing, taxiing, and discharging happy customers, who, after a short walk, picked up their luggage from a waiting carousel and walked outside to the parking lot, no more than 100 feet away. Another vignette showed a plane in the process of being unloaded, with several bins of small packages going from the plane directly into a truck, headed for a warehouse. Next, medical equipment and frozen organs were being rushed to a hospital via a waiting helicopter at the terminal. Finally, a route map was presented, along with typical travel times, which included not just the flight time, but also the time to move through the airport, pick up luggage, and reserve a rental car. Comparisons with the large hub terminals were also shown, revealing an average savings of 2 hours, door to door, for the regional carrier over flying on the major airlines.

Initially, 21 regional airports were targeted to be included in the routes. A route map showed the different combinations along with a graph portraying pertinent figures on operating costs, maintenance, insurance, taxes, and so on, of regional aircraft and the larger planes flown by the major carriers. With a projected breakeven load factor of 38% (i.e., 19 passengers out of 50), the average flight distance of 250 nautical miles (287.5 statute miles)—which took approximately 50 minutes of flying time for an average ticket price of $78.50—was competitive with the large carriers, flying from the hubs, and their shuttle operations.

A question and answer period followed, during which the public voiced its concerns. Several neighbors living within a few blocks of the airport protested any increase in flights, due to the noise and potential danger of crashes. Actually, one of the target airports had experienced a crash two years earlier when a business jet, while taking off, got caught in a high crosswind and veered off the runway during an overly aggressively applied rudder correction. The left landing gear broke off, part of which was ingested into the left engine, which began burning after takeoff. All of this caused a severe yaw, stalling the left wing, which then caused the aircraft to flip over and crash in a parking lot next to an apartment building. The four people on the plane died instantly, but at least 50 people who were occupying the apartment building at the time of the crash had no injuries.

The representatives from the airline produced a chart showing the statistics of airplane crashes, compared with those for rail, ship, and automobiles. The last mode of travel was seen as the most dangerous, by several orders of magnitude. Then another videotape featured the planned safety features of the new Aero Dash 50, Dash 100, and business jet models, including advanced designs for maintaining passenger cabin integrity, for seat and seat belt construction, and for overhead luggage bins. New impact composite materials and fabrics, impregnated with the latest fire-retardant chemicals, were to be utilized for all interior components, including galley equipment, toilets, side trim panels, and windows. A virtual reality video using the advanced heads-up display instrumentation of the new Aero Dash series, allowed participants to experience being the pilot or copilot by viewing the taxi, takeoff, final approach, and landing. Several neighbors tried it and were pleasantly surprised by the feeling of control it gave them.

The vote by the supervisors was 4 to 3 in favor of allowing Omni Central to proceed with plans to establish routes to other regional airports around the country. The situation was repeated in 14 out of 21 cities, with sufficient approvals to permit efficient route planning and profitable scheduling. The route plan had an average of three daily flights between each of the 14 medium-sized cities, either grouped as paired destinations or a chain of successive destinations.

Temporary, modular terminal buildings were to be erected on property owned by the municipalities, then replaced with regular brick and masonry structures after 2–3 years of operation. It was assumed that at least 50% of the selected sites would not develop a sufficient volume of passengers and freight to justify continued operations. Therefore, new sites were to be continuously evaluated for incorporation into the route plan.

Maintenance and fuel were contracted with existing local fixed-based operators, who shared in the passenger and freight revenues. Local travel agents, rental car agencies, and express package delivery, bus, taxi, and limousine services also were to benefit from the increases in airline traffic; in exchange for providing their respective service, each would receive a small percentage of ticket revenues in addition to the direct charge paid by the customers. Medical emergency organ transport system providers were usually compensated by either hospitals or insurance companies, with fees paid directly to the airline.

During the period of the presentations to the small city councils, Aero engineers were completing the final designs for the Dash 100, 50, and 20 series aircraft. A few problems were encountered, some considered major. The first was the determination that a 100 passenger version was too heavy to meet the 5000 ft maximum takeoff field length by at least 15,000 lb. The decision was made to resize the airframe into one designed for 70 passengers, with the Dash 50 and Dash 20 versions left intact. The fuselage cross section was left the same for all three versions, with identical overwing center section, front cockpit/galley/toilet wardrobe, and rear tail/cargo modules. Variable-length forward and rear passenger modules were designed to fit together with the modules described above. The passenger versions had four-abreast seating, with a seat pitch of 31 cm (78.74 in.), a center aisle width of 45 cm (18 in.) and a height of 191 cm (75.2 in.). This resulted in an overall inside diameter of 250 cm (98.43 in.) and an outside diameter of 260 cm (102.4 in).

The combined passenger/cargo conversion presented a challenge for the bulkhead separating the two compartments. Consultations with Omni-Central officials resulted in the consensus that only one bulkhead location had economic justification, that at the junction of the overwing and rear passenger modules. This decision simplified the design since all aircraft could be constructed with the same bulkhead fittings. Removal of rear seat assemblies was designed to be accomplished easily through the use of integrated rollers and locking pins. Seat pitch optional spacings were built in to the seat tracks. The cargo pallets were designed to utilize the same locking pin spacing sequence. The bulkhead locking mechanism utilized a method for to the interior modular connecting that was similar to that of the seat track locking pins.

Even the business jet versions could be adapted for limited cargo without compromising the executive seating arrangements. Certain interior furnishings (such as couch/bunks) would be removed if cargo capacity needed to be increased. The bulkhead design contained an optional door to permit corporate flight personnel to assist in cargo loading or unloading if desirable or advisable. Quite often, flight personnel were required to wait for executives to complete sales or business meetings, which could take 4–6 hours. This time period could be otherwise usefully occupied with express or critical cargo expediting, provided that ground personnel were also available for handling. The main reason for flight personnel involvement was to ensure that proper static weight and balance calculations were performed and proper cargo load security procedures were observed. This proposed additional workload was to be designed so as not to interfere with the flight crew's need and accommodation for rest between flights.

Other design problems encountered included utilization of a common cockpit, wing/center fuselage interface, and tail/empannage for all fuselage lengths. It was calculated in the initial sizing exercise that a common wing section could be utilized. Fuel storage, stability, and control parameters all needed to be verified with the static load and flight test prototypes, to be assembled at the completion of the design.

12.7 PROTOTYPE DEVELOPMENT

During prototype construction, several tooling-related issues were resolved, including hydroforming and bonded skin. Hydroforming, although offering great promise, was rejected. The process had already been partially validated in the automotive market, where high-volume applications on frame side rails and unitized structural components justified the large tooling expense (approximately $6.5 million per component), based on the savings in final assembly time. Bonded control panels, on the other hand, had been utilized in several production aircraft for a few years, without any incident or failure reported. The selection of the adhesive was to be based partially on the static load/fatigue, flight test results, and coupon tests from the suppliers laboratories, to be corroborated by Aero lab tests. Torsional stiffness was an extremely critical parameter, and fatigue problems had occurred from tests on earlier models. Prototype parts were fabricated utilizing a stereolithography process, in which a plastic powder material in a fluidized bed was heated in a pattern to duplicate the cross-sectional shape of the part under design.

Wing-to-fuselage attachment was also a concern. The initial sizing exercise indicated that for minimum weight, the spars should be continuous from tip to tip. However, this limited the fuel storage capacity and, therefore, range. The other concern was for crashworthiness, in the event that the outer wings were sheared off, resulting in large-scale fuel spillage and a potential fire hazard. The wing section under the fuselage was designed as a boxed socket sleeve to maintain the combination of outer wing flexibility and inner wing stiffness. The outer wing fuel cells were self sealing, as were the tanks under the fuselage. The arrangement allowed fuel to pass through the bulkheads between outer and inner wings easily, but sealed themselves in the event that a crash sheared away the outer wings.

Four different fuselage modules were fabricated, each with a unique pattern and style of longitudinal reinforcement, based on CAD outputs for improved crashworthiness. Testing would consist of a series of drop tests, whereby the section was raised vertically on a drop tower, and then allowed to free-fall and impact various surfaces. Passenger dummies, mounted in production representative seats, were strapped in properly with belts. Load cells were to record the head, neck, chest, and thigh loads, along with moments on the neck. The drop tests were also modified to permit testing of the fuel system integrity of the fuselage tanks mounted in the wing box.

Flight testing consisted of conducting takeoffs and landings, measuring field lengths, recording fuel consumption, maximum climbing and cruising speed, altitude, and range. Emergency maneuvers including collision avoidance turning, climbing, and diving were also performed. Finally, stall and spin tests were conducted, essential for any aircraft but critical for T-tailed transports.

After analyzing all test results, Aero engineers met with their management and senior executives, to draw conclusions and plan for the next phase. The static load tests were the most successful, with 259% of takeoff gross weight being applied to the wings without failure or permanent deformation. Next, the results of the fatigue tests showed that after temperature cycling, the adhesively bonded wing ribs and spar had shown some signs of film degradation. Several iterations in the adhesive formula had solved the problem adequately, since there was no major delamination, or "zippering." Surface preparation was a quality control issue at this point.

The drop test results analysis was then presented. The results demonstrated the need for additional structure to prevent collapse of the passenger cabin floor during a crash from a shallow glide angle (less than 20°). The increase in weight was estimated to be approximately 1500 lb, consisting of additional longitudinal members and cross stringers.

The thrust-specific fuel consumption at cruise was too high, to meet the 1000 NM range requirement, by approximately 12%, resulting in the lower range and cruise performance.

The flight test results showed that the empty weight of the airframe needed to be reduced by almost 10% included "resizing" the engines to a lower thrust level, which would result in longer takeoff runs, reducing payload to allow more fuel to be carried, which would extend the range, but only marginally. Reducing payload would raise the current breakeven load factor of 38% substantially, to more than 50%, requiring virtually full passenger loads in order to make a reasonable profit.

Other issues emanating from the prototype evaluations included service accessibility of some flight management hardware, a relatively low mean time between failure rate for the heads-up landing display, and sealing problems (i.e., premature failure) with the cargo doors. The conversion bulkhead panel connectors were also subject to fatigue failures. Beefed-up reinforcements were shown to improve this performance, at least in the stress lab. One of the two flight test models was retrofitted for evaluation. This latter flight test also resulted in another problem, this time with the landing gear strut, which needed to be strengthened, due to increased weight of all the "fixes."

12.8   MANUFACTURING

Many of the concerns of Aero Corp.'s manufacturing engineering department had already been consolidated into the weekly design review process. The engineers participated in the development of a Design for Manufacturing and Assembly process (DFMA), within the envelope of the overall new product development process. This system is designed to minimize the information transfer problems that can occur between the design engineering process and the manufacturing operation. The systems works by having both design and manufacturing personnel involved in the initial design processes, layout reviews, subassembly fabrication, and final assembly tryout. Besides dimensional measurement variation control between mating panels, other goals include part count reduction, manufacturing cost, and weight reduction. Commercially available software packages facilitate the processes described above.

Since the Aero regional jet fuselage consisted of four basic modules, dimensional fit-up was a critical design issue. When prototype parts were machined with information from the same computer disk that was utilized for the design, interferences or gaps between mating panels should have been minimized. However, several discrepancies were found on the outer ring mating flange that preventing complete fit-up without rework. The design engineers were called down to the prototype shop and shown the problem, which they had already found on their own, through a quality spot check, but discovered too late to head off the fabrication step. Large laid-up composite plastic interior components had to be machined to fit the mating metal panels. The prototype parts contained small defects causing interference fit problems that were delaying the first aircraft assembly. The cause of the problem was traced to the growth of the large panels from the heat of curing and machining. Application of dry ice prior to the machining process solved the growth problem. The DFMA program basically solved 95% of the problems before they even became obvious.[2]

Minor issues on the number of customized interiors and exterior colors remained as items to develop as lower-cost alternatives, if possible.

12.9   PROMOTION AND DISTRIBUTION

The ad agency that Aero contracted with to develop an advertising program had no previous experience with aircraft, which was the result of a conscious effort. Aero senior management wanted to try a completely new approach to this challenge. One of the main reasons for this decision was the need, based on research results, to attract a new type of business executive traveler. A relatively new concept of fractional ownership was to be explored for potential marketing advantages. Fractional ownership, a form of time-sharing involved leasing aircraft to a relatively large number of owners, who were issued occupational "rights" in direct proportion to their financial contribution to the aircraft. The potential advantage for the regional jet program was the flexibility of having three passenger capacities, designed for small, medium, and large groups, for charters, business trips to trade shows, overseas delegations, and other uses. Firms could tailor their travel plans to fit their immediate needs without having to purchase or lease an aircraft that might sit on the tarmac underutilized for most of the time, when it could be earning revenue for another company. Aero senior management even discussed setting up its own leasing or fractional ownership operation.

Distribution of new or used commercial aircraft was not considered a primary design issue since the company would deliver new or refurbished units directly to the airlines, with the costs amortized into the sale or lease price.

---

[2] Alfredo Herrera, The Boeing Company, conversation June 4, 1999.

12.10 AFTER-SALES SUPPORT

Service or maintenance engineering was a well-developed activity within the aerospace industry. However, this did not mean that there were no further opportunities for improvement. On the contrary, the airlines were collectively unimpressed with the manufacturer's lack of timely response on maintenance problems, which they were forced to resolve themselves. Diagnostic procedures were often exemplified as worst-case scenarios. Flight instruments frequently went out of calibration, unique service tools and kits were all too often in short supply or backordered, and engine overhaul procedure manuals were not kept up to date as often as the engine modifications, which were constantly upgraded.

Part of the market survey research done earlier in the program uncovered the above issues, and Aero senior management resolved to satisfy its customers with a new program, designed to rectify the old problems. Aero established its own maintenance training facility, not only for its only mechanics but also for airline personnel as well. Special service tools and kits were placed in an on-line expediting operation, which also provided Internet access to all diagnostic procedures for engines as well as instruments. Aero also established a hot line technical service engineering liaison office to handle all customer inquiries on a 24-hour basis.

12.11 AERO REGIONAL JET SPECIFICATIONS

Table 12.1 contains key aircraft design specificatins for the three Aero Regional Jet configurations of 70, 50 and 20 passenger capacities. Major dimensions, flight envelope data and measurements required by Federal Aviation Regulations for takeoff, landing, and noise levels have been included.

TABLE 12.1.

Aero Regional jet 50 & 70 passenger and business aircraft specifications	Mfr Model	Mfr Model	Mfr Model
Specification	Aero Dash 70	Aero Dash 50	Aero Dash 20
Wing span in meters (feet)	27.5 (90.2)	20 (65.6)	19.75 (64.78)
Wing chord @ root/tip in meters (feet)	5.25 (17.22)/1.25 (4.1)	5.25 (17.22)/1.25 (4.1)	4.15 (13.6)/1.25 (4.1)
Wing aspect ratio (span squared/gross wing area)	8.19	7.51	7.8
Wing area in square meters (square feet)	92.3 (1000)	53.25 (575)	50 (538)
Length overall/fuselage/ fuselage diameter in meters (feet)	38.0 (124.64)/ 35.25 (115.62) 2.59 (8.5)	25.5 (83.64)/ 22.75 (74.62) 2.59 (8.5)	21.35 (70)/ 19.25 (63.03) 2.59 (8.5)
Height overall/tailplane/ elevator span in meters (feet)	8.16 (26.76)/ 9.5 (31.16)	7.35 (24.11)/ 8.379 (27.48)	7.05 (23.1)/ 7.87 (25.8)
Baggage volume rear/main/ underfloor for'd/aft/ total in cubic m (ft)	3.5 (123.2)/5 (176.4)/ 8.25 (291.25)/ 4.85 (171.1)	9 (29.42)/5.35 (17.55)	4.62 (163)
Empty weight/max payload in kg (lbs)	22750 (50050)/ 9340 (20548)	13500 (29700)/ 6500 (9900)	9057 (19925)/ 3623 (7570)
Max fuel weight in kg (lbs)	4610 (10140)	3975 (8750)	3345 (7360)
Max takeoff weight in kg (lbs)	36765 (80883)	23975 (52745)	16025 (35255)
Max landing weight in kg (lbs)	34265 (75383)	21475 (47245)	13525 (29755)
Max wing loading in kg/sgm (lbs/sqft)[a]	398.32 (80.88)	450.2 (91.73)	320.5 (65.53)

[a]Jackson, Paul, et al., *Janes's All The World's Aircraft*, Jane's Information Group, Alexandria, Va., 1996. pp. 31.

(*continued*)

TABLE 12.1. (*Continued*)

Aero Regional jet 50 & 70 passenger and business aircraft specifications	Mfr Model	Mfr Model	Mfr Model
Max power loading/thrust/weight in kg/kW or kg/kN 9lb/lb st or lb/shp)	295 (2.92)	249.45 (2.44)	211.65 (2.04)
Max operating/cruising/level speed in [Mach#] [naut mi/hour] kmr/hr (mi/hr)	[.75][450] 850 (527)	[.86][340] 630 (391)	[.85][340] 477 (295)/[.82]
Approach speed/stalling speed in [naut mi/hr] km/hr (miles/hr)	[121] 219 (139)	[121] 219 (139)	[121] 219 (139)
Max rate of climb @sea/level/OEI @S/L in meters/min (feet/min)	1275 (4180)	1125 (3690)	1050 (344)
Max operating altitude/service ceiling/ceiling OEI in meters (feet)	11128 (36500)	12500 (41000)	14250 (46740)
FAR25/balanced takeoff field length/ @S/L& T-O flap in meters (feet)	1372 (4500)	1500 (4920)	1678 (5500) w/20 pass. @S/L & ISA
FAR91/25/135 lndg field @ max length landing weight @S/L ISA in m (ft)	1280 (4200)	1325 (4345)	850 (2788)
Range w/max payload @long range cruise speed [naut mi]/km (mi)	[1200] 2225 (1380)	[1000] 1402 (870)	[2750] 3857 (2391) w/20 pass & IFR res
T/O noise level in EPNdb	77.5	77.4	76.9
Approach noise level in EPNdb	89.5	90.7	90.3
Sideline noise level in EPNdb	90.3	81.3	81.7
Wheelbase in m (ft)	18.41 (60.4)	12 (39.36)	10.75 (35.26)
Wheel track in m (ft)	5 (16.4)	3.25 (10.66)	3.25 (10.66)
Number of passengers	70	50	20
Cargo/baggage door p/s f/a height width height to sill in m (ft)	p/s/a 1.4 (4.92) 1.5 (4.92)) 1.7 (5.57)	p&s/a 1.5 (4.92) 1.5 (4.92) 1.75 (5.74)	p&s/a 1.5 (4.49) 1.4 (4.92) 1.75 (5.74)

FAR25, 91/25/135 describe various Federal Aviation Regulatons for takeoff and landing for public transport aircraft with weight exceeding 5670 kg (12,500 lb).

13
HVC INC.

13.1   INTRODUCTION: A HEAVY-DUTY VEHICLE COMPANY DEVELOPS A UNIVERSAL TRACKED VEHICLE

This case study follows the development of a universal tracked vehicle designed to be adaptable to virtually any implement in the construction industry, with a minimum of conversion time and inconvenience. The case covers the 10 step new product development process, in which Simultaneous Engineering principles are applied, not only at each step, but also as early in the process as possible.

HVC Inc. had a long and illustrious history designing and producing heavy-duty and military vehicles and components, dating from the 1910s. At first, it was a major supplier of driveline components and systems to other OEMs, for early four-wheel-drive off-road trucks and some of the first military tanks utilized in World War I. Until the 1940s, HVC concentrated on four- and six-wheel drive specialized vehicles for markets such as well drilling, oil exploration, firefighting, and material hauling.

Because of their previous experience in tanks, The U.S. War Department invited HVC to participate in the development of a new generation of tank, to be competitive with the German Mk IV Panther and Russian T-34 models, that were in combat against each other in Eastern Europe. While HVC's prototypes were developed too late for World War II, their efforts resulted in several small production contracts principally for export. The HVC-19d and e models (M-45 and M-46), were involved in "brushfire" wars in Africa and the Southeast Asian theater during the 1950s and 60s.

The other company lines also prospered, as the U.S. economy expanded along with the population increases after World War II. The specialty off-road truck segments were all extremely profitable, which resulted in a very stable and sustainable average annual growth rate of 8–9% in revenue and 12–15% in gross profit. The tracked vehicle business grew at a much lower rate, but it was still important to the overall position of HVC as a niche manufacturer. Earthmoving, forestry management, and utilities services were the main customer base for this segment.

The market segments continued to subdivide until the proportion of "special or customized" to "standard" models approached 4 to 1. Costs to design and manufacture the specials were passed along to customers without any appreciable loss in market or profit. However, the influx of foreign competition resulted in less profitable business, particularly for standard units, since foreign labor costs were considerably lower than domestic rates due to traditional union contract and collective bargaining agreements. Lead times for the design, development, production, and delivery of offshore machinery was not as much of a disadvantage as was expected, because of post–World War II developments overseas.

Many of the importers facilities were recently rebuilt or brand new, having been financed through the Marshall Plan or other strategic provision. The importers also benefited from application of the most recent, innovative thinking on efficient design processes and being first to market in newer segments, with better-quality products, priced competitively or below the home market equivalents.

The domestic manufacturers initially overreacted to their foreign trade disadvantages by trying to leverage their political assets to pass tariffs on imported commercial vehicles. This ploy was only marginally effective until the importers cut their costs either through even more efficient operations or through building their own domestic plants in economically depressed areas, gaining their own political support through job creation and paying taxes. Their costs were further lowered along with lead times as the pipeline was further shortened. At this point, the domestic manufacturers were struggling for survival; they needed to dramatically change their ways of doing business or die.

13.2 STRATEGIC PLANNING

HVC found itself in these desperate straits as sales kept falling to the import/transplant heavy-duty vehicle manufacturers. A radical new approach was needed, one that would ultimately keep the company not only from bankruptcy but also growing profitably. The new business strategy called for the development of a series of standardized modules of tracked and multiwheel drive "prime movers," i.e., basic building-block units that could be combined or packaged in order to accommodate virtually any implement or device, either as a standard or special attachment. The key decision was whether to design adaptability for every possible or potential device or only into the most important or highest volume devices.

A universally adaptable wheel (tracked) prime mover would have the following as primary design variables:

Overall length, width, height, and volume

Wheelbase (maximum front centerline of longitudinal drive sprocket to rear centerline)

Wheel track (track side center/midpoint to side center/midpoint)

Overall length, center to center spacing, and height of frame (load members)

Number, mounting pad locations, and spacing of cross members (cross-frame members, side frames and end box caps)

Center-of-gravity location, maximum gross weight, and payload (implement) distribution

Ground clearance, front and rear liftover angles

Maximum tipping/rollover angle by major load (implement)

Type, location, overall size, and mounting dimensions of power plant(s)

Location of transmission(s)

Location of transfer case(s)

Location, size, and mounting pad dimensions of driving and nondriving axles and (idlers)

Size and configuration of main and secondary drive bearings and drive sprockets

Location, mounting pad dimensions, and configuration of standard/optional outriggers

Location and configuration of mounting points/pads/bases for special equipment (implements)

Location and configuration of operator's cab, safety cage, seat(s), controls and instrumentation, and display(s)

13.3   MARKET RESEARCH, IDEA GENERATION, SCREENING/EVALUATION, AND PRODUCT PLANNING

An extensive market research study was analyzed by both inside experts and outside consultants, who concluded that HVC needed to offer products that not only provided more flexibility for customers wants, but also gave the company the manufacturing flexibility to compete against lower-cost rivals. The combination of adaptability and convertibility were seen as the twin pillars of the new business strategy, a new line of prime movers for nearly all applications with sufficient volume to justify designing in special features. Brainstorming within the engineering and design departments was conducted, followed by the screening exercise.

Research clinics were conducted with major customer groups around the country, to confirm the new design philosophy. Utilizing preliminary artists' sketches and storyboards typical applications of the new designs for tracked vehicles were presented. In addition to the adaptability features of the new designs, owner/operator convertibility was also stressed. The ease with which different implements could be changed over, for example, on the tracked configuration, from scraper blade to back hoe, was highlighted. Cleanup was another advantage, where most of the tracked vehicle running gear was to be mounted inside the left and right side frames, to keep clear of debris. Hydraulic lines were to be routed inside channel covers to prevent snagging on tree roots, old power lines, and cables.

13.4   DESIGN AND ENGINEERING

The wheeled versions were also to benefit from a modular design architecture. Five-wheel-base and three-track-width variants were envisioned. Four-wheel drive was standard for the smaller three, while six-wheel drive was standard, and eight-wheel drive optional (i.e., two front steerable axles) on the larger sizes. Cab dimensions were confined to two sizes, for two and four doors and seats. Power-plant options were limited to five diesel engines, with three manual gearboxes, one automatic and one electrohydraulic, with hydraulic motors on each axle, all centrally controlled by a console computer.

During the initial design phase, both the tracked and wheeled engineering teams identified areas for improvement over the existing products in the market. Safety improvements included optional air bags for driver and passenger seating positions on the truck bodies. A padded safety cage was envisioned for the tracked model operator's seat. Seat belts were included in both designs even though rarely utilized by operators. With air bags, the front and side cab structure had to be designed to crush progressively in order to trigger deployments early enough to protect the operator and passenger. This process involved adding mass to reinforce the lower front rails and sills under the doors to prevent premature collapse from a frontal impact. Door hinges and lock mechanisms were also seen to need redesigning for added strength.

Rollover resistance was another area for strengthening, since most heavy-duty equipment in the market was potentially subjected to dangerous conditions at construction sites. One consideration was for a break-away cab, on the wheeled versions, to separate completely if the vehicle was subjected to severe side or inverted loads. For commercial vehicles, a design which added 250–300 lb to its empty weight, even for a safety issue, was not a trivial matter, since it incrementally reduced payload and profitability. Alternative designs for protection were being evaluated.

Durability concerns were also on the minds of the engineers, who had been reviewing reports from customers, complaining about premature axle bearing and seal failures, transmission shifting problems, and fuel system difficulties that plagued contractors and operators, every day. HVC engineers had conducted face-to-face interviews with maintenance technicians, who carried lists of chronic service issues, on each model, that were never seriously considered by manufacturers prior to new design reviews. The problems were simply ignored or carried over into the new product, perpetuating the condition. When the problems were considered, it was usually too late to change the drawings or to get new tests run to confirm a new approach.

The field had developed their own fixes for some of these problems, such as automatically changing bearings and seals after a few hundred hours of operation, to prevent having a failure at the job site. One diesel engine manufacturer utilized the same thread on an oil filter cap that would leak, after about 15 hours operation, due to excessive vibration in the oil filter mount. HVC management concluded that a remote mount was the best solution and specified it to the supplier, along with a more user-friendly thread for the cap.

A major plant complexity issue concerned the number of axle and bearing combinations for all available track centers and wheel bases. The total number of new part numbers was in excess of 600, including axle shafts, bearings, housings, covers, seals, and fasteners. A concerted joint effort by the suppliers, with engineering coordinating, resulted in a reduction of more than 300. This was achieved through a standardization process of reviewing each application to confirm whether a slightly larger or smaller size or performance level would suffice in the particular case. This, however, raised the weight of the total product in the applications that were upgraded. Additional machining operations were necessary in some other applications. Both processes were judged to be worthy quality improvements over earlier practices.

The only performance issue was fuel consumption. In the tracked vehicles, it was more of a fuel storage capacity problem, since most implements required some type of counterweight to balance an off-center load situation. This weight, typically a large cast iron lump, shaped to fit behind a turret or pivoting head or behind the engine bay, occupied more space than originally thought. For safety reasons, the design specified that all fuel was to be stored outside the operator compartment and the engine bay. Therefore, the only option was to store it behind the rear compartment, exactly where the counterweight was designated to go. A new process was specified in which the counterweight castings were separated into smaller pieces, allowing large cavities in the center for storage in blow-molded plastic tanks, each complete with an in-tank pump and level indicator. When the tanks were empty, the remaining weight fulfilled the minimum counter balance requirement. When full, the fuel weight was within the range of that of the maximum counterbalance requirement.

Noise was somewhat of an environmental issue, since several municipalities had passed noise ordinances, mostly affecting road vehicles. Construction equipment noise problems were usually confined to isolated complaints from residential areas near construction sites, and then only if work progressed on weekends or evenings, which were typically reserved for such activities as emergency repairs or critical fitting of support beams over roadways.

Exhaust system suppliers were contacted early, and requested to furnish ideas and prototype parts for evaluation. Severe cases were handled by multilayered metal and extensive bypass baffling networks, to break up harmonics that were propagating out from the exhaust ductwork and causing vibrations, i.e., "ringing" of the pipes and loosening fittings. This would add significantly to the power plant's cost, which needed justification, i.e., sales gained with the system relative to those lost without it. One option would allow distributors to install the system when and where required, without penalizing the base unit.

The final environmental issue was recyclability, which had traditionally been difficult to improve beyond that of the existing steel and cast-iron components. However, largely due to efforts by the affected industries in Europe, a method for dismantling wiring harnesses had been perfected to recover the copper wire. The next challenge was to extract seat back and cushion foam, after separating it from the vinyl covering. New seat designs allowed for separable inner reinforcements, instead of molding the foam around the metal parts. Vinyl covers were being designed with breakaway stitching for faster stripping, without seriously affecting wearability, similar to bags of animal feed or fertilizer.

Two chronic warranty problems were mentioned by customers: dirt ingestion into air filtration systems, and the tensioning calibration process
for tracks. The engineers created a new, tighter-fitting but more complex
inlet air duct, with baffles and dirt traps, flexible enough to fit any power
plant specified. The number and complexity of the baffles varied by application and by level of efficiency. The tensioning calibration system was
easier to solve, by switching suppliers, with a newer, more reliable, and
faster system operation.

13.5 PROTOTYPE DEVELOPMENT

Prototype testing revealed several other problems that needed to be corrected prior to production. The tracked vehicles exhibited a propensity to
throw their tracks when turning sharply or moving over extremely rough
terrain. The treads were found to be loosening prematurely, due to excessive wear in the pivot holes. The material was heat-treated, too soft, and
subject to high wear rates, particularly in abrasive conditions such as sand.
A different heat treatment corrected the wear problem but brought about
another one in its place: fatigue cracks from brittleness. A change in material was indicated, which added weight and additional manufacturing steps,
due to its toughness and wear resistance properties.

Side curtains were being considered to reduce damage from debris and
to keep mud from packing around the drive sprockets and throwing off the
tracks. Mud scrapers and deflector shields were being designed as alternatives to the curtains.

The wheeled prototype models were somewhat easier to manage, although the various suppliers of engines, axles, frames, and subsystems
made the coordination problems more prevalent than a complete in-house
design effort. There were minor problems with component fit-up and interferences, primarily with tubes, hoses, and wiring. Driveline vibration
problems resulting from the two steerable front axles proved to be more
difficult to solve. The power steering units were not balanced, resulting
in "kickbacks" during extreme or locked turns, when the front axles were
engaged. Damping shocks on the tie rod ends cured this problem.

13.6 MANUFACTURING

After analyzing the results of the prototype tests, HVC senior management
was ready to decide on the production phase. Because of the simultaneous

design process, development costs were well within the original proposals limits for potential profitability. Most of the challenges noted by manufacturing engineers early in the design phase had been addressed in the prototypes. The only remaining costs were due to the new track system and the wiring and hydraulic line modules. These operations were justified by lower projected warranty claims. The marketing department had been busy creating new advertisement copy for brochures, magazine ads, and billboards. The original themes were still valid, with no major changes in the hardware, which would have necessitated redoing all of the ad work if the family design concept had not worked.

13.7   PROMOTION AND DISTRIBUTION

Distributors were generally pleased with how the new designs came out. Their major problems resulted from making errors when filling out the order forms. Due to the complexity of earlier products, mistakes in ordering optional equipment typically caused reordering two to three times. HVC's new "family" style of products description meant that most applications were covered by one or more of the standard packages, similar to those of light-duty truck or passenger car systems. Only a few options that were not included in the packages, which was reflected in the new sales brochures.

13.8   AFTER-SALES SUPPORT

With only standard power and control modules, stocking service parts should have been an ideal situation. However, a disadvantage could occur when engineering changes had to be made to correct manufacturing problems, upgrades in parts from suppliers, or improvements in cost, weights or warranty coverage. To the extent possible, the negative impact of these changes (i.e., obsolete parts inventories, and a time lag to order and stock the new replacement parts) caused service and warrant costs to be 35% higher than the competition.

Based on the previous five years' records of service part changes, the analysis showed that the majority of changes occurred within six months of major new model introductions. The new design process resulted in fewer parts that were unique to service. This meant that regular production parts could fulfill the service obligation, resulting in design cost savings and further savings in incurring much lower service parts inventory costs and obsolescence.

14

ELECTRO-MATIC CORPORATION

14.1 INTRODUCTION: AN ELECTRONIC COMPONENTS MANUFACTURER BECOMES A SYSTEM PROVIDER

This case will follow the new product development process for a portable computer designed to permit universal compatibility with all peripheral devices that might be encountered in different modes of transportation. The process flow includes the application of Simultaneous Engineering principles to each phase, beginning with the concept phase, through market research, product planning, design and engineering, prototype development, manufacturing, marketing, and service. The Simultaneous Engineering requirements grow with additional detail as each phase in the development process is considered.

Electro-matic Corp (EMC) had grown along with the rest of the personal computer industry as a supplier of components, including microprocessors, memory chips, clocks, and so on. Prior to that, EMC had produced millions of integrated circuits (ICs) and individual electronic components for virtually all applications, including military and civilian avionics, home entertainment, appliances, motor vehicles, and so forth. Over the 60-year history of the company, the evolution from vacuum tubes and multivibrators, coils, and variable condensers to very large scale integrated circuits had been accompanied by the explosive growth of a global mass production manufacturing industry, which had doubled in size every 5–7 years in spite of recessions, wars, and political turmoil on an unprecedented scale, much of it in undeveloped or underdeveloped countries.

Along with the high production came lower profits, as increased competition between companies and countries devastated pricing policies in every market. EMC had long since moved its component fabrication operations overseas, constantly seeking to remain even with or stay ahead of their competition with regards to manufacturing costs. The VLSI production had remained a domestic operation only because of the need to maintain as short a communication lead time as possible between the design freeze point and production orders. However, EMC had finally come to a point where there was no place left to go for cheaper labor, as the costs to ship the parts in and assemblies out and to not only build the plant but assist with building up an infrastructure to support it—including electric power generation, worker recruitment, training transportation and even lodging—could not be recovered in the market, on either a short- or near-term basis.

Worldwide industrial development had also begun to be scrutinized for causing irreversible environmental problems, due to failure to protect the local and regional area, when the plants were shut down after the few, if any, profitable years. Lack of cleanup of toxic wastes, acids, heavy metals, and other contaminants utilized in the production of circuit boards was the major problem, resulting in effluent discharged into streams, which poisoned groundwater aquifers. The local governments were neither strong nor stable enough to prevent factory owners from simply walking away from the mess, leaving destitute workers to deal with the long term unemployment, and medical and other health and physiological issues. Many underdeveloped countries yearned for the hard currency that such plants would bring, along with new schools, hospitals, and so on. They tended to overlook the long-term problems, i.e., if the plants were not successful. However, past problems were catching up with these countries and with the manufacturers, who had avoided these issues far too long.

14.2   STRATEGIC PLANNING

Senior management at EMC had decided, in their annual strategic planning meeting, to explore the possibility of developing their own personal computer to interface with all currently available and advanced peripherals and input channels, including Internet and satellite communications. The main goal of this proposal was to create a new product with greater brand value potential than in the existing component parts group, but still to stay with the company's expertise. They felt that an integrated approach might be more valuable to customers, who could then design a system around the EMC computer regardless of environment or intended usage.

One concept concentrated on the consumer carrying the EMC PC to a vehicle (car, truck, airplane, train, or boat) and then attaching it to a port replicator/docking station permanently installed in the instrument panel or seat back. The PC could be installed in a console below the seat cushion. Utilizing a series of voice commands, the user could "boot up" the computer and open programs as usual. Public or semi-public transportation could have the computer locking equipment built into the seat backs, along with speakers, display screen, mouse, and other peripherals. Printing, faxing and e-mail could be queued up through a central server, on board or at a remote location.

14.3   MARKET RESEARCH AND IDEA GENERATION

The concept was developed into a workable or demonstrable "breadboard" unit and shown to customers, along with a number of peripheral setups for different transport vehicle applications. It was constructed as though the vehicles each had terminal screen, Internet connection, built-in cellular phone, fax, and printer/plotter, all voice and infrared activated. Separate peripherals, along with a detachable flat display panel and stand, would also be optional for users who envisioned working or utilizing the equipment away from the office or transportation unit.

Market research surveys on customers' experiences and complaints with existing PCs had been cataloged for future reference. Incompatibility between competitive operating systems generated the largest number of complaints. Lack of uniformity of program keys, redundant keys, (e.g., Print Screen, Scroll Lock, Pause), difficult or awkward mouse controls, inconsistent on-off switch locations, ambiguous and tedious initialization (boot-up and shut-down) procedures, and modular designs that are not easily upgradable for memory represent some of the other high response concerns.

Several recommendations also resulted from the survey, including more accessibility to "open source " software, "plug and play" peripherals, a more compact, lighter, but also more robust and shock-resistant case, a briefcase-like handle, access doors on the side for loading additional memory packs, and launch buttons on top for accessing common program applications.[1]

[1] Walter S. Mossberg, "Steve Jobs Makes Bold Leap in PC Design and Advances an Inch," *The Wall Street Journal*, January 4, 1999, pp. B4.

14.4   SCREENING/EVALUATION AND PRODUCT PLANNING

Designers were also aware of several problems that were not visible to users but remained concerns nonetheless. The most serious included electromagnetic interference (EMI) with adjacent devices, battery drain/power consumption, and excessive heat buildup, a potentially life-shortening phenomenon for sensitive electronic components.

Aside from designing and fabricating the ICs, EMC engineers were concerned about adding features to achieve higher levels of standardization, and presumably, lower manufacturing costs, when research showed that most users did not take advantage of all features on a regular basis. At the same time, they felt uncomfortable with specifying that customers could upgrade their base-level units later, without a visit to the factory service center. The track record for customers following instructions in software installation packages had not been very favorable. Complicated installation and recalibration of the system did not seem to be a likely scenario. Poorly written and misinterpreted documentation was also part of the problem.

As previously mentioned, safety was the main concern of the EMI protection system. The concern was not over the ability to design the protection apparatus, but rather, what would happen to the transportation operational systems if a failure in the protection mechanism occurred? This was probably a worst-case scenario, but a valid concern nonetheless. Early designs utilized "mu" metal shielding inside the cases of EMI-prone circuits. It was thought that a more active design might be more cost- and mass-effective, without complicating the assembly process significantly.

A nontechnical safety issue was discussed: the moral or ethical dilemma faced when entertainment or business-related information displays distract the driver who then subsequently causes an accident. At first, EMC senior management refused to acknowledge the existence of the potential liability problem, until company lawyers made a presentation on the risks associated with ignoring it. Finally, senior managers admitted their own shortsightedness and committed to design interlocks that would restrict entertainment programs to rear seat locations and included shielding to keep the driver from seeing the front screen when the vehicle was being driven. Information for navigation using the global positioning satellite (GPS) system and for emergency situations (i.e., when receiving a signal from an ambulance or fire truck) was to be presented in a heads-up display that projected the warning signals, data, or navigational instructions, during driving, on the inside of the windshield, around the periphery of the driver's eye's central focusing ellipse. Shielding for direct sun or oncoming headlamp glare would need to be developed.

14.5   DESIGN AND ENGINEERING

Aside from designing and building the ICs, the problems of integrating the universal port adapter or docking station for the PC presented a series of special concerns. The vehicle operating environment was not considered friendly compared with that of an office, plant, or home. It included temperature extremes, vibration, shock from rough road conditions, contamination from spilled fluids, food particles, and corrosion from being left in vehicles with carpets saturated from road salt. Computers typically are exposed to many of these conditions, particularly in commercial or military vehicles. However, regularly scheduled maintenance tends to limit the severity of the consequences. Military equipment also tends to be more robust than civilian appliances.

Personal use computers generally tend to be treated like any other appliance. The main concern expressed by EMC engineers was that a defective PC could corrupt or damage the vehicle's host computer systems, potentially endangering the occupants if a failure could shut down a safety, ignition, or fuel control system. Designing the proper "fire walls" to insulate the vehicle from a disruptive computer signal, from either direction, represented a significant design initiative and workload impact, considering that this was a relatively new input.

14.6   PROTOTYPE DEVELOPMENT

The breadboard prototypes were well received by major and potential customers, who tried out different programs during a variety of mobility situations. Feedback included concern over the ability of the heads-up display to be viewed accurately during bright daylight, vibration problems with some of the plug-and-play modules (e.g., CD players and zip drives), limited hard drive space, mouse defects, radio static related to the PC's power supply, and a case of the screen "jitters" (a flashing or line jumping that seemed to occur independently of any particular vehicle motion, attitude, or environmental exposure). All of the comments were collected for future design and development, typically two to three years after initial production. EMC's president surprised the engineering team when he demanded that all "defects" or observations be purged from the design and all corrections incorporated into the first production units assembled for sale.

14.7 MANUFACTURING, PROMOTION, DISTRIBUTION, AND AFTER-SALES SUPPORT

Early production units suffered from the usual defects in IC connections, glitches in wire-wound post connections, open circuits in plated-through holes, incorrect terminal strips and connectors from suppliers, and so on. All these defects were considered acceptable compared with past new product introductions. This introduction was, however, different because the first 1000 units were road tested for several hours, both in real vehicles and on rough road simulators, in order to identify any loose connections or other defects.

The testing also gave the marketing and advertising departments sufficient material to create a campaign that showed genuine differentiation from the other PC-based systems in existence. Plans for selling peripheral products got a similar opportunity when the separate displays and keyboards were subjected to the same tests as the vehicles and test rigs.

After-sales marketing created a schedule whereby customers would be able to have free memory pack and hard drive upgrades whenever they chose or when EMC service technicians and the customers both agreed on a convenient time and place, including service calls at home or office. This plan put the technician on a personal contact basis with the customer, for not only hardware but software problems as well.

15

XYZ PACKAGING CORPORATION

15.1 INTRODUCTION: A PACKAGING COMPANY DEVELOPS PACKAGING DESIGN GUIDELINES FOR NEW PRODUCTS

This case study is concerned chiefly with the development of guidelines for the packaging of consumer goods. The goal of Simultaneous Engineering is to achieve recognition for the value of getting these recommendations into the hands and minds of new consumer product developers as early in the process as possible, and to refine them as the new product development process continues.

XYZ Packaging was founded just after World War II through the merger of three separate entities, an advertising agency specializing in consumer goods, a product design studio, and a company that manufactured large packaging equipment, such as folding cardboard boxes, flat shirt carton material, paper file folders, along with other paper office supplies. Company executives developed a strategy whereby they could offer a complete range of services related to product packaging and distribution that would be more cost-effective than manufacturers could do themselves or through multiple suppliers. XYZ developed business with virtually all of the manufacturers of cereals and other breakfast foods, pharmaceuticals, toys, laundry, paper, and other products selling through supermarkets. The business grew rapidly for more than 25 years.

The postwar economic growth, in step with the expanding population, resulted in the promotion of the supermarket industry. Prepackaged foods, toys, hardware, and other convenience items were competing for shelf space, as stores strived to keep up with a seemingly insatiable demand for more variety and better-quality products.

Most manufacturers had developed their packaging operations over many years, with relatively stable production requirements and long product life cycles. It was not uncommon for a package to remain the same size, shape, and color, and to have the same graphics for 15–20 years. Shelf life was not a particular concern, nor were date codes, bar codes, or "appeal" (i.e., awareness) packaging. XYZ had prospered because the market was changing faster than the manufacturers could adapt. XYZ offered an alternative; by developing the packaging concepts alongside the new product, production costs and time to market were both dramatically reduced. Their packaging machinery designs were flexible, so that container size and shape could be updated quickly and easily. Printing, colors and label generation equipment were also designed for optimizing change.

Countering the growth phenomenon was a trend toward more individualized products and customized services, with less of a "discard" approach. Environmental concerns had resulted in an increasing awareness, particularly in the 1970s, about groundwater and air pollution, roadside litter, and landfill space running out. Political pressure was brought to bear on the consumer goods industries to adopt more environmentally friendly packaging, which would enable more voluntary conservation of resources and encourage recycling of packaging materials.

While this trend was slowly making headway, an incident sparked a new concern about product safety. Bottles of over-the-counter drug products were contaminated with poison, resulting in death or serious internal injury to several dozen people in scattered areas. Authorities spent months analyzing the relationship between the product manufacturing processes and the packaging operations before they deciphered that tampering had occurred in the stores, instead of at an earlier phase in the manufacturing process. The packaging industry reacted quickly to protect the public with tamper-proof caps, plastic seals, and vacuum-sealed cans and bottles for food and drugs.

The resulting trade-off between safety and convenience had another consequence—complaints about difficulties opening packages. One of the more recent survey analyses concluded that more than 60% of consumers had experienced problems with opening tamper-proof packages. They ranged from the inability to tear a perforated cardboard seal, to finding a place to open a shrink-wrapped container, to difficulty in opening a drug bottle cap, among other problems. In future designs, it was assumed that solving these problems would result in new marketing opportunities.

15.2   STRATEGIC PLANNING, MARKET RESEARCH, IDEA GENERATION, SCREENING/EVALUATION, AND PRODUCT PLANNING

XYZ executives also interviewed their counterparts at the product manufacturers. Several new opportunities were discussed, including more colorful and demonstrative packaging, more brand awareness, instructional information, additional space needed for labeling, bar code reading, and export market requirements. As a result, XYZ developed a new strategy to exploit these new ideas. Their research laboratory had been perfecting a new clear-plastic bubble wrap that permitted shrink-wrapping tightly around nonuniformly shaped objects. The quality of the final version had exceptional optical clarity, consistent thickness, and stability through the normal temperature range. A few challenges remained, including compatibility with the adhesives commonly utilized to bond the plastic to cardboard backing sheets without causing separation in the cardboard. Other issues included finding a somewhat prominent yet inconspicuous location for the recycling identification mark, the right combination of additives to prevent discoloration or yellowing of the plastic, minimizing warping from high humidity and temperature extremes, and insuring against the lack of contamination from contact with food.

15.3   SIMULTANEOUS ENGINEERING

XYZ's philosophy called for their principle's designers and engineers to get involved early in new product programs so that packaging considerations could be discussed and decided on. The process consisted of reviewing the new product concept, evaluating the brand image, determining whether there was adequate space for the "message," analyzing the packaging concept, i.e., how to open the outer package, retain or dispose of the outer package, and final disposition of the inner package, if necessary. Recycling concerns were also part of the discussion, with the potential recyclability ranked for each proposed packaging material. Toxicity and compatibility of adhesives and other issues were also covered, along with shipping and handling constraints, i.e., fragility, combined product and package weight, nesting or other potential risks of product damage during shipping, and labeling issues.

Finally, the discussion centered on potential export considerations, including brand transfer, language translation problems (i.e., making sure no offensive phraseology or ambiguous language was contained on the package or in the instructions), and whether the package itself was acceptable to the distribution channels available. For example, some food stores in rural countries might not display products on racks, but have them setting on shelves. Therefore, an irregularly shaped shrink-wrapped item, normally hung from a wire rack, would have to be repackaged to be placed vertically upright in a shelf, or inside a small rectangular box, permitting it to be stacked with others.

15.4   NEW PRODUCT PACKAGING GUIDELINES

In order to facilitate the whole process, XYZ staff members had prepared a checklist of packaging guidelines for product development engineers at the manufacturing companies to utilize when new products were being planned. The checklist was organized to generally follow the same logic as the new product process steps seen earlier. The checklist is highlighted below:

Strategic Considerations

Consistent fit with existing products, if applicable
Consistent fit with brand image, if applicable
Consistent fit with brand or product "message," if applicable

Safety Considerations

Potential Product Contamination From:

Air
Moisture
Pressure change
Temperature change
Natural sunlight
Indoor light
Noise
Package material toxicity/degradation

Potential Product Damage From:

Crush
Bend

Stretch
Twist
Shock

Convenience Issues (Customer Jury Consensus)

Instructions:

Clear
Consistent
Practical

Opening:

Clean
Consistent
Practical

Reseal:

Clear
Consistent
Practical

Recycle (Package):

Separable by material
Practical

Recycle (Product):

Separable by material
Practical

Labeling

Brand "message":

Intact
Consistent with other products with same brand

Bar Code:

Readable

Date Code:

Readable

Recycling Mark:

Readable

Export Mark(s):

Readable

Foreign language check

Package Design

Outer Container:

Folded paper box
Shrink wrap plastic
Combination

Adhesive Usage:

Compatibility with package
Compatibility with product
Separability

Inner Container (if Required):

Plastic
Metal
Paper
Composite
Separability

Shipping Container:

Cardboard
Plastic tub
Wooden/plastic pallet
Recyclable

Design Cost Parameters

Safety (security from tampering, shred-resistant paper)

Durability (package seal integrity, adhesive strength, and aging resistance)

Environmental Compliance (raw material and finished goods emissions and toxicity)

Functional Performance (formability, shrink-wrap ratio, bondability, etc.)

Recyclability (thermoplastic melt temperature, separability from other materials)

Quality (optical clarity, ink run resistance, folding flap conformance, etc.)

Damageability (scratch, dent, and cut resistance)

Warranty (defect identification and customer mistakes)

SECTION III

ADVANCED PRODUCT DESIGN AND DEVELOPMENT TECHNOLOGIES

III.1 INTRODUCTION

This section of the book is devoted to several new concepts for the utilization of Simultaneous Engineering for new product development, including developing parametric design optimization models, conducting virtual design practices, and addressing new or unanticipated parameters. These concepts are either already in use or being planned for introduction. They are designed to facilitate the overall design process and reduce design iteration and/or decision time. They accomplish this by considering all of the possible factors that could influence product appearance, function, interaction with upstream and/or downstream operations, customers, suppliers, regulations, and so forth, as early in the product's life as possible, and by continuously reviewing these influences at each step of the new product development process. Identification of all parameters and their relative ranking may vary by product line, development process phase element, industry, and market.

Optimization of all design parameters simultaneously involves understanding how the parameters relate to one another over the spectra of each parameter's value range, from best to average to worst-case scenarios. Precise knowledge of the interrelationships is probably not necessary if the general or trend relationship is known or suspected. The model development is based on obtaining, inferring, or interpolating the data and allowing the program to develop the interrelationships, based on the knowledge breadth and depth of the user. Then the optimization (or minimization) of cost parameters can be performed.

Chapters 16–19 all seek to develop a multivariate optimization model by comparing minimum values of design parameter-based dependent variables for a given range of independent variables that represent the designers' or engineers' best estimate for the general area where all of the minimum values reside. Chapter 16 is the automotive model, following, in general, the discussion in Chapter 11. Chapter 17 is the aerospace model, aligned with Chapter 12. Chapter 18 is for heavy-duty or off-road commercial equipment, and corresponds to Chapter 13. Chapter 19 looks at the electronics model, and Chapter 20 looks to the future. *Note:* A detailed parametric analysis of consumer packaged goods is considered to be beyond the scope of this text, based on the complexity and variability of the products involved. However, general issues relating to the packaging constraints have been included in this chapter.

III.2   COMMON PARAMETERS ACROSS DIFFERENT INDUSTRIES

The five industries under study have separate parametric relationships, which must be developed into cost-based or other-based equations before optimization can take place. However, there are similarities between some industries, between automotive/light vehicle, military/heavy-duty vehicle, electronics, and aerospace. For example, mass or weight minimization may be similar in both automotive and aerospace. Reliability and durability may be approached similarly between automotive, aerospace, electronics and heavy-duty vehicle designs for testing, with length and severity of the testing as variables. Some programs are combined when electronic devices, common on board all vehicles now, are tested along with other systems on the vehicles through accelerated durability tests, which expose the vehicle to such extremes as temperature, humidity, vibration/fatigue, shock, and corrosion. Other design parameters also have importance and are of common interest across different industries, such as environmental impact, energy consumption, functional performance, manufacturability, serviceability, maintainability, repairability, and recyclability. Some parameters that are unique to a particular industry can have a dramatic effect on overall design intent and execution.

Crashworthiness should have the highest priority, for obvious reasons, on new designs of light-, medium-, and heavy-duty vehicles, buses, rail passenger cars, and aircraft. Safety also has a high ranking in packaged goods with childproof seals and date codes for proper shelf life tracking. Electronic units need proper handling while being attached to 120 volt

A.C. service for operation or battery charging, in spite of typically having low power consumption.

Electromagnetic interference (EMI) is extremely important in the electronics industry, because it can not only affect the operation of a singular unit, such as a portable or laptop computer, but also influence peripheral equipment, such as aircraft onboard navigation and communications equipment.

Toxicity of materials affects the recyclability potential of products, if manual dismantling procedures are involved, in vehicles' electronic components and consumer goods packaging.

Application adaptability or convertibility is certainly important in the aircraft and heavy vehicle markets, where configuration complexity can have a direct bearing on the profitability of operations. Idle equipment due to unscheduled maintenance or conversion to a different set of customer requirements is not earning revenue for the business. Equipment upgradability versus replacement issues can have short-term economic implications for electronic equipment, when determining whether future capabilities should be designed into a new product even when current usage of the advanced function may be several years ahead.

All of the above are product design parameters that are dependent or influenced by other less dependent or by independent parameters. For example, crashworthiness is a direct function of vehicle mass. Since the energy absorption capabilities of materials and construction methods are the result of reactions to the kinetic energy imparted to them (the product of mass and one half the velocity squared), mass is a key factor in controlling how energy is dissipated in a structure during a collision or impact. The test speed is regulated, so it is the mass that must be changed if the vehicle's crashworthiness is to be improved, which usually means a reduction in the mass through a change in material (and/or gauge thickness).

If mass reductions are not a feasible solution to solve a crashworthiness problem, another potential solution may be to improve the force deflection properties of primary structural energy-absorbing members through a shape or configuration change. By adding buckling beads, dimples, holes, or reinforcements to the front rails, to stiffen a given cross section, the crash pulse to trigger the passive restraint system could be modified to suit the need that originally arose. This action, which increases the number of stamping operations and adds to the manufacturing costs, is referred to as complexity. It includes all factors that add directly to the cost of the manufacturing process—such welding, casting, machining, fabrication, and assembly—of individual component parts, subsystems, systems, labor, paint, coatings, fasteners and other attaching hardware, and so on. In some cases, both mass and complexity are added, such as with reinforcements, which add mass and have to be stamped, punched, and welded in the rail, either inside or outside.

Mass and complexity, typically but not always, are both added to vehicles to improve durability and reliability performance. Structural components on light-duty vehicles often need to be beefed up, or upgauged, in order to survive accelerated durability testing, if the original design exhibited fatigue cracks or unusual wear characteristics. However, with the advent of computer-aided engineering (CAE) programs, high-stress conditions can be eliminated without adding mass, by changing the shape or configuration of the component while still on the screen. In aircraft and heavy-duty vehicles, durability problems are often solved by enlarging drive axles, bearings, and their respective housings. Aircraft landing gear have to withstand large *g* forces, while wings and control surfaces withstand aerodynamic forces and moments. When abnormal fatigue cracks appear in structural elements, on flight test aircraft, a more robust design is needed. Most of the time, redesigns of this type can be heavier and more complicated to manufacture.

The same philosophy can be extended into the environmental compliance activity, where vehicle mass determines the settings on the chassis dynamometer that is utilized to simulate rolling road resistance and aerodynamic drag. If vehicle mass exceeds a certain limit, the dynamometer is calibrated to the next higher setting, which requires the vehicle to expend more fuel and emits more exhaust gases, all of which are regulated. Failure to pass an emissions test for certification will most certainly result in intense mass reduction activities and additional product cost for lighter-weight material substitutions, elimination of product content, and other changes. This could also have a direct bearing on the marketability of the vehicle, particularly if fuel consumption costs are a prime factor in the cost of ownership.

Fuel consumption is analyzed differently for aircraft, where it directly affects operating costs, profitability, and ticket sales promotions, among other factors. It is a direct function of the aircraft mass and the weight of the fuel, passengers, crew, and luggage. Fuel consumption is a performance parameter, which can determine optimum range, altitude, and payload capacity.

Power consumption for electronic equipment should be evaluated in the same light as fuel consumption for transportation vehicles. Excessive consumption robs these devices of a portion of their utility, whether battery or alternating current powered. In the end, it is an environmental impact issue, as equally important as emissions and independent of where the fuel comes from or the intermediate factors involved with its conversion. Ultimately, it boils down to using up a scarce resource, and that issue must be addressed, even if there is limited regulatory control or no competitive pressure to do so.

The other dependent parameters have similar effects on the five industries that have been highlighted. The ranking or priority of the parameters may change, but the ultimate goal of simultaneous optimization should remain constant. Also, new parameters may become prominent or the priority within a given product line may shift as a result of market changes and customer demands. Provisions for handling new inputs and changes need to be made in such a way that there is minimal or no disruption in the design process for a new product. The process for documenting and executing change is not well established, because the circumstances under which it is a necessity are not well thought out either. However, once the optimization process is in place and functioning well enough that the participants trust its output, modifying it for new requirements may not be a monumental task. The key is that there has to be trust by everyone that the right decisions will be made, more timely with the benefit of technology than without it. This premise is based on senior management having good communications and interpersonal skills so that discussions on design alternatives can be conducted in a framework of cooperation and openness.

Much has been written about the barriers still existing that prohibit effective communication between different departments, such as marketing, purchasing, engineering, manufacturing, and service. Protection of responsibilities, "turf battles," fear of loss, ego massaging, and other motivations are the reasons why these actions persist in business today, even after all that has been said, written, and done to alleviate this hindrance to progress. The future may hold the key to solving this problem, if and when the new technology proposed here is allowed to create a higher level of interaction and trust.

16

AUTOMOTIVE OPTIMIZATION MODEL

16.1 INTRODUCTION: HOW PARAMETRIC AUTOMOTIVE DESIGN FUNCTIONS CAN BE OPTIMIZED

For the automotive/light-duty vehicle industry, eight dependent and two independent parameters have been identified as being directly or indirectly design sensitive. The list could be twice or half as long depending on specific product requirements. The interrelationships also depend on product specific issues. Each manufacturer will have to evaluate how its own experiences relate the new product to the historical base of previous designs.

Each of the eight dependent cost parameters have been plotted versus the two independent parameters, manufacturing complexity and mass reduction costs. The resultant composite dependent parameter is the total direct manufacturing cost. The other dependent contributors include the following parameters:

Structural compliance (for meeting safety standards)

Environmental compliance (for meeting emissions and corporate fuel economy regulations)

Vehicle functional performance objectives (for improving acceleration, braking, handling, etc.)

Manufacturing quality objectives (for improving customer satisfaction indices for paint, body fit, finish, NVH, squeaks, water leaks, etc.)

Serviceability and warranty targets (for improving electromechanical component accessibility and disassembly/reassembly without disturbing other components)

Durability and reliability targets (for improving fatigue life of body, trim, and mechanical and electrical components)

Damageability/repairability objectives (for lowering insurability risks of both collision and comprehensive coverage, theft being the largest portion)

Recyclability objectives (for increasing the number and type of metallic and nonmetallic components and subsystems that can be recycled)

The key to building any mathematical model is the ability to analytically describe the type of physical behavior encountered or outcome desired. The development of a transfer function that represents some action/reaction, measurable/quantifiable relationship may be easier to visualize than putting the resultant in the form of a cost parameter. This latter conversion requires a different approach, whereby opportunity costs, based on estimating the most likely, best, and worst-case outcomes, have to be identified and estimated. There is also the concept of unanticipated and future parameters yet to be identified or quantified. It may be possible to estimate the short-, near-, and long-term effects of future regulations, particularly safety related, if the company has been developing new technology for compliance. Relating the new parameter to an existing one may be another viable strategy.

The graphs that follow show the general relationship between the eight parameters identified earlier. Some vary directly with both independent variables, and others vary inversely. The goal is to optimize or minimize all eight at once by finding a common solution or the answer that comes the closest to meeting all eight equations. Once the relationships have been established, a matrix of equations with eight dependent and two independent variables can be constructed. Such a set of equations has been developed for the graphical solutions provided later in the chapter. This somewhat simplistic case is necessary to allow for adequate explanation in the text and so a three-dimensional graphical solution can be shown in two dimensions. It should also be noted that the software package utilized in this exercise, MATLAB, can also find minimum values for all dependent variables, as long as they all have the same number of terms and are generally similar in value. Then the minimum value for all dependent variables, in the realm of the independent parameters, can also be plotted, so that the sensitivity of the relationships can be studied.

16.2   PARAMETRIC INTERACTIONS

The independent parameters are mass reduction cost and manufacturing complexity cost. The x-axis is mass reduction cost, the y-axis is the manufacturing complexity cost, and the z-axis is the product cost of the dependent parameter under analysis. The number of dependent parameters plotted range from two to four, depending on the particular graph being viewed. The equations relating these variables contain many terms both directly and inversely related to one another. For the purpose of explanation, it is assumed that all eight dependent parameters have a nonzero minimum value for some nonzero value combination of x and y. The minima also vary by different values of x and y. The dependent parameters are initially weighted equally, but this situation could be varied if another weighting was deemed significant. The issue of prioritization could provide the impetus to change the weighting. For example, if both structural safety and environmental compliance were to be ranked higher or more heavily than the others or an alternative scheme, then the model could accommodate it by scaling the z parameters higher for those items. How the design bureau or engineering office ranks the parameters for a given new product will depend not only on their relative ranking, but also on how far away the parameters are from their target values (i.e., the ones farthest away might be weighted the highest, and therefore, have the most effect on the total cost of the product).

For the sake of simplicity, in the example, in the Appendixes, each of the eight z parameters was assumed to vary from 0 to 1200, while both x and y parameters vary from 0 to 700. Each z function has a minimum value somewhere in the 0–700 range for both x and y. Some have their minima relatively close to the origin, while for others it occurs farther away.

Starting with the assumption that the new vehicle is all new and not derived from an earlier platform, an analysis of potential interactions needs to be performed. A typical scenario is given to illustrate what might occur in the current design atmosphere and how Simultaneous Engineering could provide a more efficient and cost-effective alternative. The following example shows how parametric priority selection can affect project outcomes.

Compliance with safety standards and emission regulations have the highest priority, while the other six parameters have met their preliminary targets for mass and complexity. The initial CAD layout of the front body structure was designed for minimum mass and relatively uncomplicated assembly, with no intermediate rail seams between the bumper mounting area and the dash panel/front floor pan joint. Subsequent simulations for

35 mph front crash response showed that force deflection of the front rails was unacceptable, with deformation propagating all the way to the dash panel. The initial reaction of the body structures group was to increase the gauge of the high-strength steel in the base structure and add several internal and external reinforcements, to initiate crush farther forward in the rail. This option would add 175 lb, or approximately $87.50, to the entire front-end sheet metal subassembly and would add significantly to the complexity cost of the assembly (about $430). Other costs may include increased collision repair expenses, ($550–1100/vehicle, due to increased complexity for removal and reinstallation of the heavier-gauge reinforced areas of the structural components). Potential problems could also occur in the area of quality: as the number of welds increases to accommodate the additional pieces, the possibility of defective welds causing additional flexure or other handling problems during turning also increases. The primary result might be an increased risk of not meeting the emissions and fuel economy regulations due to the added mass, unless an equivalent amount of mass can be reduced through other initiatives.

Since the optimization process model is primarily configured to evaluate potential design alternatives, utilization is predicated on having the following facilities available: a database of direct costs for each major component group, system, or subsystem, estimated by all of the dependent variables defined above, in terms of mass reduction and/or manufacturing complexity costs.

The net effect of having several widely varying functions, all intersecting one another in the same space, is fairly representative of the conflicts and trade-offs typically encountered in the current design environment. Trying to find values of x and y (i.e., mass reduction and manufacturing complexity) that satisfy of all values of z (i.e., engineering parameters) that minimize manufacturing cost while meeting all of the product design requirements is at the heart of using Simultaneous Engineering within the new product development process. Having all the pertinent data at hand should facilitate design decisions, if the data are given a proper hearing, and if nontechnical issues do not overshadow the technical content of the design review setting.

For each of the eight dependent variables it has been assumed that a matrix of different mass reduction and manufacturing complexity combinations exist that will provide a minimum value for each z parameter. Assuming also that the x and y values, for all minimum z values, are never going to be the same, one solution is to combine the eight equations and solve for a minimum value of z for all eight. For simplicity in explaining the model, each of the eight functions has the same number and type of terms. All functions were created from fictitious, but reasonable, data and have very pronounced minimum values of both x and y. This process also assumed that all would have the same weighting or priority. However, this can be varied easily by changing the function coefficients before combining the functions. This same logic is followed throughout the text, utilizing the MATLAB software function *meshgrid*.[1] Table 16.1 (at the end of the chapter) shows a typical example of the *meshgrid* command sequence (i.e., for four surfaces).

16.3   GRAPHICAL SOLUTIONS

For the purposes of explanation, it will be assumed that each dependent design parameter varies quadratically with both of the independent variables, mass reduction cost and manufacturing complexity cost. This means that initial increases in mass reduction and complexity may produce a small improvement in the design parameter in question, further increments would produce proportionately larger improvements, until at some point, the incremental improvements decrease, resulting in increasing costs, due to diminishing returns. Graphically, the curves are carpet plots, of at least two dependent parameters displayed against the two independent parameters, mass reduction and manufacturing complexity cost, with the minima for each parameter located at a specific x, y, z coordinate set (shown as black dots). The aggregate of all eight parameters has a similar shape, along with minimum values of mass reduction and manufacturing complexity costs.

Several combinations of parameters have been paired together in Figures 16.1–16.10 to show how the minima vary across the range of mass reduction and manufacturing complexity functions. For the purpose of simplification, only two functions are shown on the same plot.

Figure 16.10 is the total product cost, the sum of all eight parameters, weighted equally. In most of these cases, solving for the minimum values of each parameter would result in 16 different values for mass reduction and complexity and 8 for the design parameters. The plots show the minimum values as well as the general shape of the functions. If all design parameters are weighted equally, one set of values will occur. If they are weighted unequally, another set of values will result. The model has sufficient flexibility to accommodate these kinds of variations. For example, Figure 16.11 shows four parameters—safety, performance, repairability, and durability—plotted simultaneously so that the minima for all except

[1] *MATLAB*, put out by The Math Works, Inc., 24 Prime Park Way, Natick, MA, 01760.

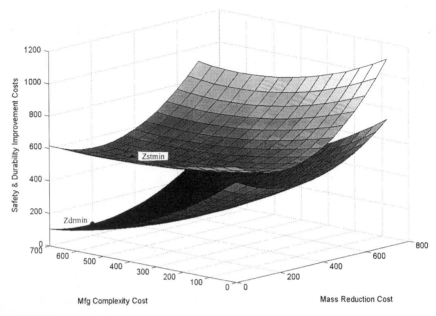

FIGURE 16.1 Safety and Durability Improvement Costs vs. Mass Reduction and Manufacturing Complexity Costs

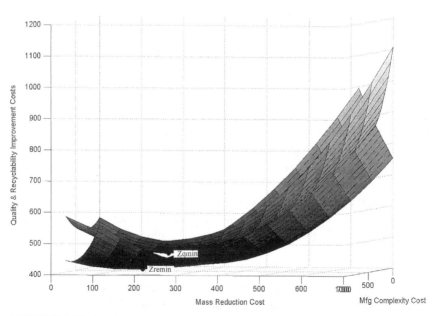

FIGURE 16.2 Quality and Recyclability Improvement Costs vs. Mass Reduction and Manufacturing Complexity Costs

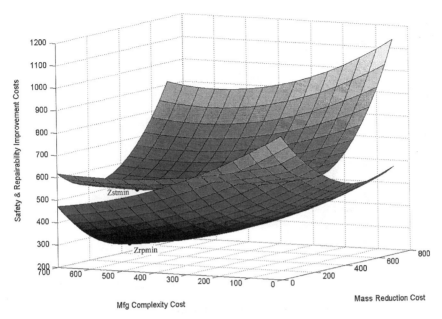

FIGURE 16.3 Safety and Repairability Improvement Costs vs. Mass Reduction and Manufacturing Complexity Costs

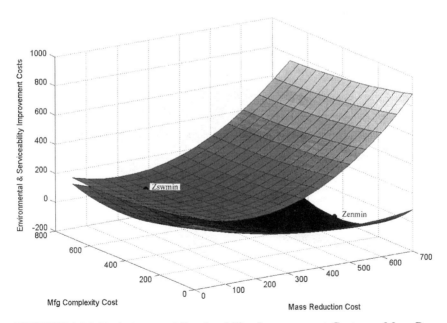

FIGURE 16.4 Emissions and Serviceability Improvement Costs vs. Mass Reduction and Manufacturing Complexity Costs

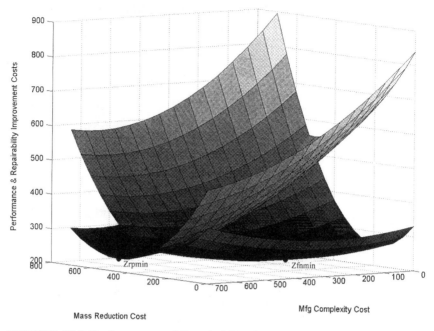

FIGURE 16.5 Performance and Repairability Improvement Costs vs. Mass Reduction and Manufacturing Complexity Costs

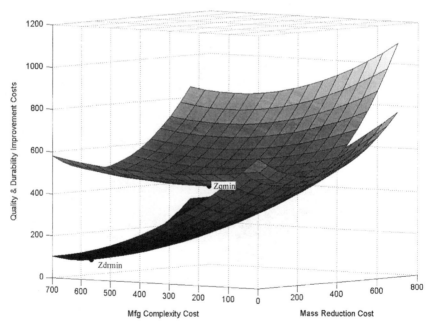

FIGURE 16.6 Quality and Durability Improvement Costs vs. Mass Reduction and Manufacturing Complexity Costs

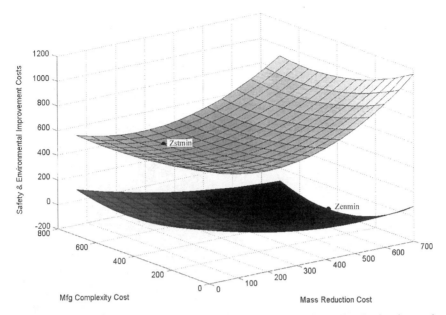

FIGURE 16.7 Safety and Emissions Improvement Costs vs. Mass Reduction and Manufacturing Complexity Costs

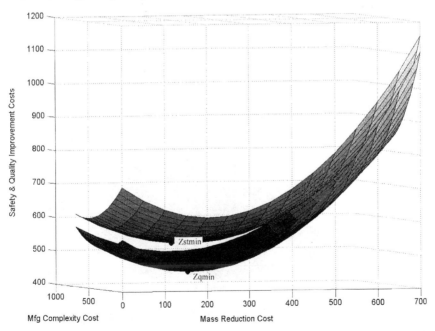

FIGURE 16.8 Safety and Quality Improvement Costs vs. Mass Reduction and Manufacturing Complexity Costs

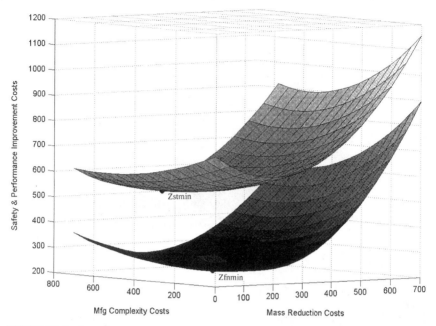

FIGURE 16.9 Safety and Performance Improvement Costs vs. Mass Reduction and Manufacturing Complexity Costs

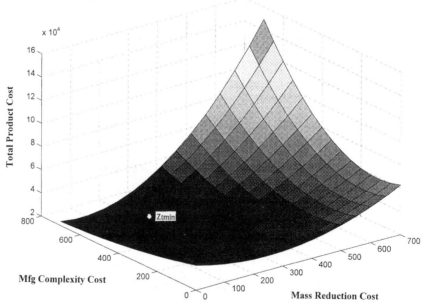

FIGURE 16.10 Total Product Cost vs. Mass Reduction and Manufacturing Complexity Costs

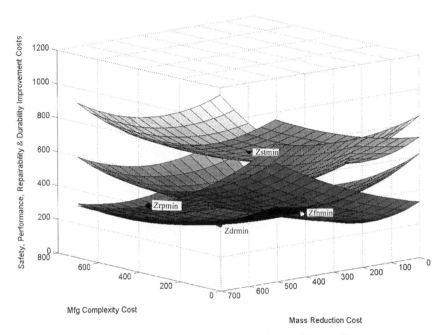

FIGURE 16.11 Safety, Performance, Repairability, and Durability Improvement Costs vs. Mass Reduction and Manufacturing Complexity Costs

repairability are visible. The other four parameters are plotted together, as shown in Figure 16.12. Plotting more than four dependent parameters is not feasible, since the overlap of the surfaces would preclude identification of all minimum values, with as many as four hidden from view, depending on the viewing angle. Combining the functions, represented as trial 49_var, is shown in Figure 16.10, with the combined effects of all eight parameters.

Calculation of the parameters utilizing the MATLAB function *fmins*, which can be utilized to find minima from several variables simultaneously. The sequence to calculate *fmins* for the eight parameters individually (Zst, Zq, Zfn, Zrp, Zen, Zre, Zdr Zsw) and the sum of all eight (Zt), is shown in Table 16.2. The index in Table 16.3 can be utilized to identify the separate and combined trials for both *fmins* and *meshgrid* programs. (Tables 16.2 and 16.3 are found at the end of the chapter.)

It can be seen that, in nearly all of the above trials, except for the aggregate, Zt, the *x* and *y* minima are different. This implies a compromise decision for the values of mass reduction and manufacturing complexity. While not ideal, it certainly represents a measure of reality. Nontechnical

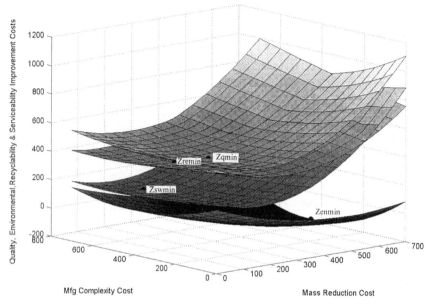

FIGURE 16.12 Quality, Emissions, Recyclability, and Serviceability Improvement Costs vs. Mass Reduction and Manufacturing Complexity Costs

issues will also play an important role. The importance of the data is the establishment of boundaries within which decisions based on data analysis can be made instead of through hearsay or guesswork.

Future requirements can be estimated in a similar manner, as long as the relationship between the dependent and independent variables can be estimated.

16.4 REFERENCE TABLES

The following Tables 16.1 and 16.2 are the results of the MATLAB calculations for the Meshgrid, the plotting program that gives the minimum values for each z parameter and the Fmins programs, for the minimum values of x and y, that correspond to the z parametric functions. Table 16.3 presents the relationships between the functions and the various plots.

TABLE 16.1. MATLAB Meshgrid Sample Command Sequence (for Four Surfaces: Automotive)

```
>> x_range=0:50:700;
>> y_range=0:50:700;
>> [X Y]=meshgrid(x_range,y_range);
>> Z1=(.0027.*(Y./2).^2-1.5088.*(Y./2)+790.0912).*(X./600).^2-
    (.0013.*(Y./2).^2-.7501.*(Y./2)+540.6235).*(X./600)+
    .0014.*(Y./2).^2-.7619.*(Y./2)+712.9559;
>> Z2=(.0026.*(Y./2).^2-1.6879.*(Y./2)+829.7853).*(X./600).^2-
    (.0013.*(Y./2).^2-.5461.*(Y./2)+484.0794).*(X./600)+
    .0013.*(Y./2).^2-.3596.*(Y./2)+330.5618;
>> Z3=(.0013.*(Y./2).^2-0.3895.*(Y./2)+394.6235).*(X./600).^2-
    (.0007.*(Y./2).^2-.3718.*(Y./2)+672.4412).*(X./600)+
    .0021.*(Y./2).^2-1.7809.*(Y./2)+839.2412;
>> Z4=(.002.*(Y./2).^2-1.244.*(Y./2)+529.8853).*(X./600).^2-
    (.0027.*(Y./2).^2-1.7051.*(Y./2)+476.4853).*(X./600)+
    .0033.*(Y./2).^2-2.6261.*(Y./2)+617.8294;
>> surf(X,Y,Z1);
>> hold on
>> surf(x,Y,Z2);
>> surf(X,Y,Z3);
>> surf(X,Y,Z4);
>> Xlabel('Mass Reduction Cost');
>> Ylabel('Mfg Complexity Cost');
>> title('Safety, Performance, Repairability & Durability
    Improvement Costs vs Mass Reduction & Mfg Complexity Costs');
>> Min_Z1=min(Z1(:));
>> Min_Z2=min(Z2(:));
>> Min_Z3=min(Z3(:));
>> Min_Z4=min(Z4(:));
>> [z1_row, z1_col]=find(Z1==Min_Z1);
>> [z2_row, z2_col]=find(Z2==Min_Z2);
>> [z3_row, z3_col]=find(Z3==Min_Z3);
>> [z4_row, z4_col]=find(Z4==Min_Z4);
>> min_x1=x_range(z1_col);min_y1=y_range(z1_row);
>> min_x2=x_range(z2_col);min_y2=y_range(z2_row);
>> min_x3=x_range(z3_col);min_y3=y_range(z3_row);
>> min_x4=x_range(z4_col);min_y4=y_range(z4_row);
>> plot3(min_x1,min_y1,Min_Z1,'go','MarkerFaceColor','g')
>> plot3(min_x2,min_y2,Min_Z2,'ro','MarkerFaceColor','r')
>> plot3(min_x3,min_y3,Min_Z3,'bo','MarkerFaceColor','b')
>> plot3(min_x4,min_y4,Min_Z4,'yo','MarkerFaceColor','y')
>> colormap gray
>> plotedit
>> ROTATE3d
```

TABLE 16.2.   MATLAB Fmins program example

When solving the fmins function, the first term after a = { } is Xabmin, the second is Yabmin. To find the equivalent Zabmin, locate Xmin and Ymin on the appropriate curve and read Zabmin on the vertical axis.

```
>> v=[0 0];
>> a=fmins('trial21_var',v)

a=

  Xmin=224.0809   Ymin=536.1441

>> v=[0 0];
>> a=fmins('trial22_var',v)

a =

  Xmin=203.5906   Ymin=308.5225

>> v=[0 0];
>> a=fmins('trial23_var',v)

a =

  Xmin=205.9495   Ymin=319.8151

>> v=[0 0];
>> a=fmins('trial24_var',v)

a =

  Xmin=438.6090   Ymin=752.1946

>> v=[0 0];
>> a=fmins('trial25_var',v)

a =

  Xmin=933.4806   Ymin=460.0713

>> v=[0 0];
>> a=fmins('trial26_var',v)

a =

  Xmin=155.7103   Ymin=383.0461

>> v=[0 0];
>> a=fmins('trial27_var',v)

a =

  Xmin=196.7162   Ymin=836.8765

>> v=[0 0];
>> a=fmins('trial28_var',v)
```

TABLE 16.2. (*Continued*)

```
a =

  Xmin=154.1126   Ymin=527.1554
>> v=[0 0];
>> a=fmins('trial29_var',v)
a =

  Xmin=243.6564   Ymin=558.1121
```

Trial 21_var

```
function b=two_var(v);
X=v(1);
Y=v(2);
b=(.0027.*(Y./2).^2-1.5088.*(Y./2)+790.0912).*(X./600).^2-
   (.0013.*(Y./2).^2-.7501.*(Y./2)+540.6235).*(X./600)+
   .0014.*(Y./2).^2-.7619.*(Y./2)+712.9559;
%%%%%%%%%%%%%%%%%%%%%%%%%%%%%end of function
```

Trial 22_var

```
function b=two_var(v);
X=v(1);
Y=v(2);
b=(.0026.*(Y./2).^2-1.7087.*(Y./2)+834.8676).*(X./600).^2-
   (.0011.*(Y./2).^2-.6239.*(Y./2)+499.7471).*(X./600)+
   .0014.*(Y./2).^2-.4241.*(Y./2)+556.0235;
%%%%%%%%%%%%%%%%%%%%%%%%%%%%%end of function
```

Trial 23_var

```
function b=two_var(v);
X=v(1);
Y=v(2);
b=(.0026.*(Y./2).^2-1.6879.*(Y./2)+829.7853).*(X./600).^2-
   (.0013.*(Y./2).^2-.5461.*(Y./2)+484.0794).*(X./600)+
   .0013.*(Y./2).^2-.3596.*(Y./2)+330.5618;
%%%%%%%%%%%%%%%%%%%%%%%%%%%%%end of function
```

(*continued*)

TABLE 16.2. (*Continued*)

Trial 24_var

```
function b=two_var(v);
X=v(1);
Y=v(2);
b=(.0013.*(Y./2).^2-0.3895.*(Y./2)+394.6235).*(X./600).^2-
   (.0007.*(Y./2).^2-.3718.*(Y./2)+672.4412).*(X./600)+
   .0021.*(Y./2).^2-1.7809.*(Y./2)+839.2412;
%%%%%%%%%%%%%%%%%%%%%%%%%%%%%%end of function
```

Trial 25_var

```
function b=two_var(v);
X=v(1);
Y=v(2);
b=(.0026.*(Y./2).^2-1.3174.*(Y./2)+300.4647).*(X./600).^2-
   (.0009.*(Y./2).^2-.5176.*(Y./2)+491.5029).*(X./600)+
   .0031.*(Y./2).^2-1.2939.*(Y./2)+249.6206;
%%%%%%%%%%%%%%%%%%%%%%%%%%%%%%end of function
```

Trial 26_var

```
function b=two_var(v);
X=v(1);
Y=v(2);
b=(.0014.*(Y./2).^2-0.6202.*(Y./2)+643.9059).*(X./600).^2-
   (.0025.*(Y./2).^2-1.8504.*(Y./2)+561.9029).*(X./600)+
   .0013.*(Y./2).^2-.724.*(Y./2)+532.2118;
%%%%%%%%%%%%%%%%%%%%%%%%%%%%%%end of function
```

Trial 27_var

```
function b=two_var(v);
X=v(1);
Y=v(2);
b=(.002.*(Y./2).^2-1.244.*(Y./2)+529.8853).*(X./600).^2-
   (.0027.*(Y./2).^2-1.7051.*(Y./2)+476.4853).*(X./600)+
   .0033.*(Y./2).^2-2.6261.*(Y./2)+617.8294;
%%%%%%%%%%%%%%%%%%%%%%%%%%%%%%end of function
```

TABLE 16.2. (*Continued*)

Trial 28_var

```
function b=two_var(v);
X=v(1);
Y=v(2);
b=(.0011.*(Y./2).^2-0.7449.*(Y./2)+852.5647).*(X./600).^2-
  (.0016.*(Y./2).^2-0.8454.*(Y./2)+488.0382).*(X./600)+
  .0012.*(Y./2).^2-.6222.*(Y./2)+291.5088;
%%%%%%%%%%%%%%%%%%%%%%%%%%%%%end of function
```

Trial 29_var

```
function b=two_var(v);
X=v(1);
Y=v(2);
b=(.0163.*(Y./2).^2-9.2214.*(Y./2)+5176.188).*(X./600).^2-
  (.0121.*(Y./2).^2-7.2104.*(Y./2)+4214.821).*(X./600)+
  .0151.*(Y./2).^2-8.5927.*(Y./2)+4129.953;
%%%%%%%%%%%%%%%%%%%%%%%%%%%%%end of function
```

TABLE 16.3. Index to MATLAB Functions and Plot References

Dependent Parameter	Parameter Description	Function	Case	Trial No.	Figure No.	*fmins* Table No.
Zst	Safety	meshgrid fmins	Auto "	21 21	16.1, 16.3, 16.7, 16.8, 16.9, 16.11	16.2
Zq	Quality	meshgrid fmins	" "	22 22	16.2, 16.6, 16.8, 16.12	16.2
Zfn	Performance	meshgrid fmins	" "	23 23	16.5, 16.9, 16.11	16.2
Zrp	Repairability	meshgrid fmins	" "	24 24	16.3, 16.5, 16.11	16.2
Zen	Environmental	meshgrid fmins	" "	25 25	16.4, 16.7, 16.12	16.2
Zre	Recyclability	meshgrid fmins	" "	26 26	16.2, 16.12	16.2
Zdr	Durability	meshgrid fmins	" "	27 27	16.1, 16.6, 16.11	16.2
Zsw	Serviceability	meshgrid fmins	" "	28 28	16.4, 16.12	16.2
Zt	Total product	meshgrid fmins	" "	29 29	16.10	16.2

17

AEROSPACE MODEL DEVELOPMENT

17.1 INTRODUCTION: HOW PARAMETRIC AIRCRAFT DESIGN FUNCTIONS CAN BE OPTIMIZED

The analytical model for the Aero Corp. Regional Jet program is an exercise in refining the design for each of the three seating configurations for 70, 50, and 20 passengers. Common themes are developed across each configuration, such as fuselage cross section, engine/nacelle combination wing/fuselage mounting, control surfaces, and tail plane. Wing loading may be held constant if weight and balance work out to be at an optimum. The basic sizing calculations are assumed to have been completed, with refinement, based on the results of the parametric study, forthcoming. As in the automotive case, all of the design parameters can either be weighted the same or differently depending on the priority assigned. In most cases, safety or crashworthiness should have the highest priority. Calculations can be difficult when assigning a value to crash survivability, but utilization of actuarial tables can establish levels of lost income from a principal family member. Recent jury awards have established a pattern for pain and suffering from nonfatal injuries, and mental anguish of survivors of the crash and relatives of the victims. All of the above are academic when reality sets in. The best suggestion for weighting for crash survivability is to multiply the safety parameters by a quantity that will ensure that safety has the absolute highest priority.

17.2 PARAMETRIC DEFINITIONS

Based on the estimated manufacturing cost of the 70-passenger model (Aero Dash 70), of $10.4 million, mass reduction and manufacturing complexity cost target limits of approximately $10,000, or approximately 0.1% of the total production cost, were initially imposed on each dependent design cost parameter. The parameters being considered were defined as follows; some are similar to automotive in that they represent incremental costs to improve the original design but also result in increased mass reduction and manufacturing complexity costs:

Safety and Crashworthiness Cost: Additional reinforcement in passenger cabin section stringers and floor separation from fuselage to wing mount; stronger seat supports (i.e., legs and brackets), seat tracks, and track reinforcements.

Environmental Compliance Cost: Redesigned engine bypass routing and tail cone to reduce takeoff noise (local noise ordinances).

Functional Performance Expectations: Reduced mass to meet empty weight for greater range and lower takeoff field length.

Quality Improvements: Redesigned interior trim and seating modules for improved comfort and eye appeal; heavier and more costly fabrics for side panels, seat backs, and cushions; thicker seat and back foam biscuits.

Durability Improvements (Based on Fatigue Test Results): Stronger fuselage outer wall stringers, module end reinforcements, tail cone, control surface and landing gear mount reinforcements, engine nacelle–wing mount reinforcements, galley rack and cabinet mount gussets, and so on.

Repairability Improvements: Repairable modules in landing gear hydraulic systems, open mount circuit boards in all vendor-supplied "black boxes," ground support consoles, engine overhaul kits, auxiliary power unit (APU), overhaul kits—all requested by the airlines and business jet leasing/fractional ownership companies.

Serviceability, (Maintainability), Improvements: Provide additional access doors to avionics bays, under instrument panel, for testing/problem diagnosis or removal/reinstallation of black-box items, without disturbing cockpit area; additional fasteners and cover plates for hydraulic landing gear system and for improved servicing of internal O-rings and seals; larger access panel and compartment for APU in the tail cone, due to the need for a larger APU.

Recyclability Challenges: Seat foam cushion and back "buns" completely separable from covers and seat cushion and back "pans";

fiberglass-reinforced control surfaces completely separable from aluminum reinforcements and mounting brackets, hardware, etc.; strippable adhesives utilized throughout fuselage and wing subassemblies.

17.3 GRAPHICAL SOLUTIONS

The following plots shows a hypothetical solution that minimizes all of the above functional parameters. One assumption could simply have both mass or weight reduction and manufacturing complexity costs added incrementally, increasing from the base design configuration. However, this special case provides only a trivial solution; i.e., minimum mass and complexity occur at the origin, meaning that nothing is added to the original design. While this is a compromise of sorts, it achieves none of the targets represented by the additional operations listed. A more realistic approach is to assume, as in the case of the automotive situation, that all of the functions have minima for both mass and complexity away from the origin, and sufficiently different from each other that some graphical or calculated compromise is in order. Then an intelligent decision can be made in selecting the combination of increased mass reduction and complexity costs that comes the closest to satisfying all eight conditions shown above.

Figure 17.1 shows a three-dimensional representation of four of the eight dependent parameters—safety, performance, repairability, and durability improvement costs—plotted against the independent parameters mass reduction and manufacturing complexity costs. Zfnmin, Zdrmin, and Zrpmin are visible as black dots near the bottom of the curves, respectively, while Zstmin lies near the middle of the Zst surface. Table 17.1 provides an index to the plots and the parameters. Table 17.2 lists the minima for the above parameters as well as the total product cost from all eight parameters. This was provided due to the inability to show all eight minima graphically on the same plot. Figure 17.2 shows three of the four other minima—Zqmin, Zremin, and Zswmin—grouped closely together, approximately at the point where Zq, Zre, and Zw surfaces intersect. Zenmin is visible near the bottom of the Zen surface, quite far from the other minima. Figure 17.3 represents the total product cost based on the net effect of all eight parametric functions, with minimum values for x and y and all parameters weighted equally. The value of Ztmin is an estimate based on the value of x and y from the fmins calculation using the trial 49 function. The limitations of the Meshgrid software preclude plotting two functions whose x, y, and z limits are not matched (i.e., the limits for Zt are approximately $10\times$ those of the individual z functions. Note that

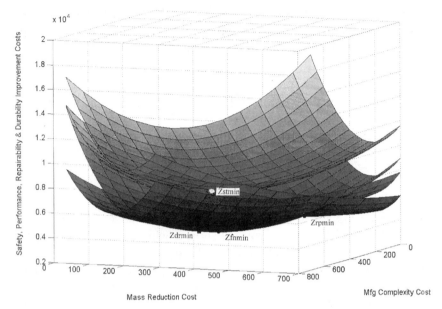

FIGURE 17.1 Safety, Performance, Repairability, and Durability Improvement Costs vs. Mass Reduction and Manufacturing Complexity Costs

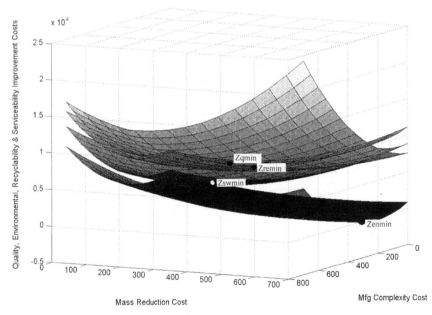

FIGURE 17.2 Quality, Environmental, Recyclability, and Serviceability Improvement Costs vs. Mass Reduction and Manufacturing Complexity Costs

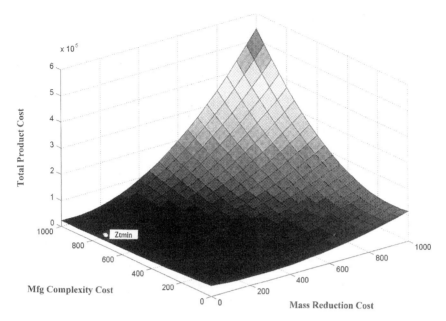

FIGURE 17.3 Total Product Cost vs. Mass Reduction and Manufacturing Complexity Costs

this does not necessarily mean that all eight are optimized individually. It just means that all of the data can have an equal analysis, and while new "what if" scenarios may always be in the background, based on all of the available technical data, intelligent and timely decisions on new product designs can only be made better.

TABLE 17.1. MATLAB Meshgrid Sample Command Sequence (for Four Surfaces: Aerospace Vehicles)

```
>> x_range=0:50:700;
>> y_range=0:50:700;
>> [X Y]=meshgrid(x_range,y_range);
>> Z1=(0.1.*(Y./2).^2-15.1.*(Y./2)+7900.9).*(X./600).^2-
   (0.1.*(Y./2).^2-7.5.*(Y./2)+5406.2).*(X./600)+
   0.1.*(Y./2).^2-7.6.*(Y./2)+7129.6;
>> Z2=(0.1.*(Y./2).^2-16.9.*(Y./2)+8297.9).*(X./600).^2-
   (0.1.*(Y./2).^2-5.5.*(Y./2)+4843.5).*(X./600)+
   0.1.*(Y./2).^2-3.6.*(Y./2)+3305.6;
>> Z3=(0.1.*(Y./2).^2-3.9.*(Y./2)+3946.2).*(X./600).^2-
   (0.1.*(Y./2).^2-3.7.*(Y./2)+6724.4).*(X./600)+
   0.1.*(Y./2).^2-17.8.*(Y./2)+8392.4;
>> Z4=(0.1.*(Y./2).^2-12.4.*(Y./2)+5298.9).*(X./600).^2-
   (0.1.*(Y./2).^2-17.1.*(Y./2)+4764.9).*(X./600)+
   0.1.*(Y./2).^2-26.3.*(Y./2)+6178.3;
>> surf(X,Y,Z1);
>> hold on
>> surf(x,Y,Z2);
>> surf(X,Y,Z3);
>> surf(X,Y,Z4);
>> Xlabel('Mass Reduction Cost');
>> Ylabel('Mfg Complexity Cost');
>> title('Safety, Performance, Repairability & Durability
   Improvement Costs vs Mass Reduction & Mfg Complexity Costs');
>> Min_Z1=min(Z1(:));
>> Min_Z2=min(Z2(:));
>> Min_Z3=min(Z3(:));
>> Min_Z4=min(Z4(:));
>> [z1_row, z1_col]=find(Z1==Min_Z1);
>> [z2_row, z2_col]=find(Z2==Min_Z2);
>> [z3_row, z3_col]=find(Z3==Min_Z3);
>> [z4_row, z4_col]=find(Z4==Min_Z4);
>> min_x1=x_range(z1_col);min_y1=y_range(z1_row);
>> min_x2=x_range(z2_col);min_y2=y_range(z2_row);
>> min_x3=x_range(z3_col);min_y3=y_range(z3_row);
>> min_x4=x_range(z4_col);min_y4=y_range(z4_row);
>> plot3(min_x1,min_y1,Min_Z1,'go','MarkerFaceColor','g')
>> plot3(min_x2,min_y2,Min_Z2,'ro','MarkerFaceColor','r')
>> plot3(min_x3,min_y3,Min_Z3,'bo','MarkerFaceColor','b')
>> plot3(min_x4,min_y4,Min_Z4,'yo','MarkerFaceColor','y')
>> colormap gray
>> plotedit
>> ROTATE3d
```

TABLE 17.2. Index to MATLAB Functions and Plot References

Dependent Parameter	Parameter Description	Function	Case	Trial No.	Figure No.	*fmins* Table No.
Zst	Safety	meshgrid	Aero	41	17.1	17.2
		fmins	"	41		
Zq	Quality	meshgrid	"	42	17.2	17.2
		fmins	"	42		
Zfn	Performance	meshgrid	"	43	17.1	17.2
		fmins	"	43		
Zrp	Repairability	meshgrid	"	44	17.1	17.2
		fmins	"	44		
Zen	Environmental	meshgrid	"	45	17.2	17.2
		fmins	"	45		
Zre	Recyclability	meshgrid	"	46	17.2	17.2
		fmins	"	46		
Zdr	Durability	meshgrid	"	47	17.1	17.2
		fmins	"	47		
Zsw	Serviceability	meshgrid	"	48	17.2	17.2
		fmins	"	48		
Zt	Total product	meshgrid	"	49	17.3	17.2
		fmins	"	49		

TABLE 17.3. MATLAB Fmins Program Example

When solving the fmins function, the first term after a = { } is Xabmin, the second is Yabmin. To find the equivalent Zabmin, locate Xmin and Ymin on the appropriate curve and read Zabmin on the vertical axis.

```
>> v=[0 0];
>> a=fmins('trial41_var',v)

a=

  Xmin=212.8569  Ymin=88.7013

>> v=[0 0];
>> a=fmins('trial42_var',v)

a =

  Xmin=183.9149  Ymin=49.9650

>> v=[0 0];
>> a=fmins('trial43_var',v)

a =

  Xmin=179.4646  Ymin=43.8646

>> v=[0 0];
>> a=fmins('trial44_var',v)

a =

  Xmin=482.3417  Ymin=205.9219

>> v=[0 0];
>> a=fmins('trial45_var',v)

a =

  Xmin=603.2212  Ymin=209.0138

>> v=[0 0];
>> a=fmins('trial46_var',v)

a =

  Xmin=260.4240  Ymin=4.4845

>> v=[0 0];
>> a=fmins('trial47_var',v)

a =

  Xmin=235.1697  Ymin=282.3069

>> v=[0 0];
>> a=fmins('trial48_var',v)

a =

  Xmin=168.6389  Ymin=55.2845

>> v=[0 0];
>> a=fmins('trial49_var',v)

a =

  Xmin=60.3835  Ymin=810.6265
```

TABLE 17.3. (*Continued*)

Trial 41_var

```
function b=two_var(v);
X=v(1);
Y=v(2);
b=(0.1.*(Y./2).^2-15.1.*(Y./2)+7900.9).*(X./600).^2-
  (0.1.*(Y./2).^2-7.5.*(Y./2)+5406.2).*(X./600)+
  0.1.*(Y./2).^2-7.6.*(Y./2)+7129.6;
%%%%%%%%%%%%%%%%%%%%%%%%%%%%%end of function
```

Trial 42_var

```
function b=two_var(v);
X=v(1);
Y=v(2);
b=(0.1.*(Y./2).^2-17.4.*(Y./2)+8373.3).*(X./600).^2-
  (0.1.*(Y./2).^2-6.2.*(Y./2)+4997.5).*(X./600)+
  0.1.*(Y./2).^2-4.2.*(Y./2)+5560.2;
%%%%%%%%%%%%%%%%%%%%%%%%%%%%%end of function
```

Trial 43_var

```
function b=two_var(v);
X=v(1);
Y=v(2);
b=(0.1.*(Y./2).^2-16.9.*(Y./2)+8297.9).*(X./600).^2-
  (0.1.*(Y./2).^2-5.5.*(Y./2)+4843.5).*(X./600)+
  0.1.*(Y./2).^2-3.6.*(Y./2)+3305.6;
%%%%%%%%%%%%%%%%%%%%%%%%%%%%%end of function
```

Trial 44_var

```
function b=two_var(v);
X=v(1);
Y=v(2);
b=(0.1.*(Y./2).^2-3.9.*(Y./2)+3946.2).*(X./600).^2-
  (0.1.*(Y./2).^2-3.7.*(Y./2)+6724.4).*(X./600)+
  0.1.*(Y./2).^2-17.8.*(Y./2)+8392.4;
%%%%%%%%%%%%%%%%%%%%%%%%%%%%%end of function
```

Trial 45_var

```
function b=two_var(v);
X=v(1);
Y=v(2);
b=(0.1.*(Y./2).^2-13.2.*(Y./2)+3004.6).*(X./600).^2-
  (0.1.*(Y./2).^2-5.2.*(Y./2)+4915).*(X./600)+
  0.1.*(Y./2).^2-12.9.*(Y./2)+2496.2;
%%%%%%%%%%%%%%%%%%%%%%%%%%%%%%end of function
```

(*continued*)

TABLE 17.3. (*Continued*)

Trial 46_var

```
function b=two_var(v);
X=v(1);
Y=v(2);
b=(0.1.*(Y./2).^2-6.2.*(Y./2)+6439.1).*(X./600).^2-
   (0.1.*(Y./2).^2-18.5.*(Y./2)+5619).*(X./600)+
   0.1.*(Y./2).^2-7.2.*(Y./2)+5322.1;
%%%%%%%%%%%%%%%%%%%%%%%%%%%%%%%end of function
```

Trial 47_var

```
function b=two_var(v);
X=v(1);
Y=v(2);
b=(0.1.*(Y./2).^2-12.4.*(Y./2)+5298.9).*(X./600).^2-
   (0.1.*(Y./2).^2-17.1.*(Y./2)+4764.9).*(X./600)+
   0.1.*(Y./2).^2-26.3.*(Y./2)+6178.3;
%%%%%%%%%%%%%%%%%%%%%%%%%%%%%%%end of function
```

Trial 48_var

```
function b=two_var(v);
X=v(1);
Y=v(2);
b=(0.1.*(Y./2).^2-7.6.*(Y./2)+8533.6).*(X./600).^2-
   (0.1.*(Y./2).^2-8.5.*(Y./2)+4880.4).*(X./600)+
   0.1.*(Y./2).^2-6.2.*(Y./2)+2915.1;
%%%%%%%%%%%%%%%%%%%%%%%%%%%%%%%end of function
```

Trial 49_var

```
function b=two_var(v);
X=v(1);
Y=v(2);
b=(0.8.*(Y./2).^2-92.7.*(Y./2)+51794.5).*(X./600).^2-
   (0.1.*(Y./2).^2-72.2.*(Y./2)+42150.9).*(X./600)+
   0.1.*(Y./2).^2-85.8.*(Y./2)+41299.5;
%%%%%%%%%%%%%%%%%%%%%%%%%%%%%%%end of function
```

18

HEAVY-DUTY/OFF-ROAD/ MILITARY VEHICLE MATHEMATICAL MODEL DEVELOPMENT

18.1   INTRODUCTION: HOW PARAMETRIC HEAVY-DUTY VEHICLE DESIGN FUNCTIONS CAN BE OPTIMIZED

This area will concentrate on the design trade-offs present in the discussion contained in Chapter 13 on standard versus customized track vehicle "prime movers." Recalling the parametric targets for the major design tasks, the maximum range of additional mass and manufacturing complexity were both in the magnitude of $600, or approximately 0.05182% of the average base price of the unit ($964,850). This worked out to be roughly 2080 lb of weight would have to be reduced to meet all of the criteria described in the tasks in Chapter 13. The additional weight would detrimentally affect powertrain durability, fuel capacity, payload, and could impose center-of-gravity restrictions on implements. The key objective was to meet as many of the parametric targets at minimum additional mass and manufacturing complexity costs. The challenge was to find the minimum combination of mass reduction and complexity costs that would minimize all of the design parameters as well. Each parameter, when plotted, was a surface comprised of various combinations of mass and complexity costs. The intersection of all these surfaces at the point closest to the origin would give both independent parameters as minima. This can be calculated and shown graphically as well. The design parameter cost targets have been listed below:

Safety (of operator/driver): Additional reinforcement around operator station, for improved energy absorption; reinforcements for seat pedestal swivel mount; additional protection around vital computer control modules, hydraulic lines, and hose fittings, for protection against toxic fluid spills and fire.

Environmental Compliance: Noise control of diesel exhaust to be controlled through sound baffle attenuation techniques, involving multi-chambered tubes.

Functional Performance Improvement: Fuel consumption increases due to extra weight to be counteracted through larger fuel storage areas outside in protected storage tanks.

Quality Improvements: Laser-cut steel blanks, and electrical discharge machining (EDM) of mounting holes, for superior fit-up; additional interior brackets, reinforcements, and tie-down fasteners and brackets, for controlling noise and vibration.

Durability Improvements: Side curtains to protect tracks from mud buildup; shielding and bracketry for hydraulic lines and fittings.

Repairability Improvements: Repairable modules in hydraulic systems and fittings; larger track treads, links, pins, and spreaders; additional access panels for automatic transmission governor and shift band adjustment, minor engine repairs, and so on; more access panels, for electronic module removal and reinstallation.

Recyclability Improvements: All plastic components, marked and recyclable; wiring harness (less connectors).

Serviceability (Maintainability) and Improvement: Reinforced brackets, hardware, and new design oil filter; shielding for hydraulic lines, hoses, fittings, wiring harnesses, etc. Field-repairable wiring harnesses (detachable connectors, circuit breakers and switches); field-accessible circuit boards for electronic modules.

18.2 GRAPHICAL SOLUTIONS

Calculated and graphical solutions for optimizing the above eight design parameters, simultaneously, for minimum mass and manufacturing complexity cost. Figures 18.1–18.3, 18.4 and 18.5 represent various combinations of parameters, with minima shown as black or white dots. This is due to the constraints of the MATLAB Meshgrid program, which plots multiple curved surfaces, with upper surfaces "shading" lower surfaces.

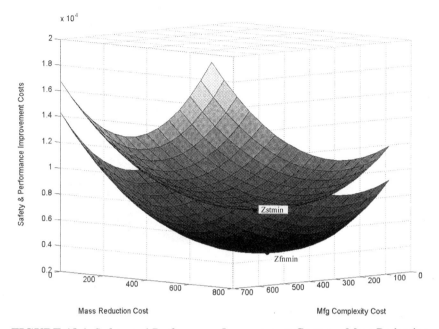

FIGURE 18.1 Safety and Performance Improvement Costs vs. Mass Reduction and Manufacturing Complexity Costs

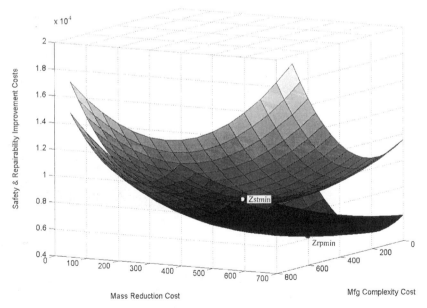

FIGURE 18.2 Safety and Repairability Improvement Costs vs. Mass Reduction and Manufacturing Complexity Costs

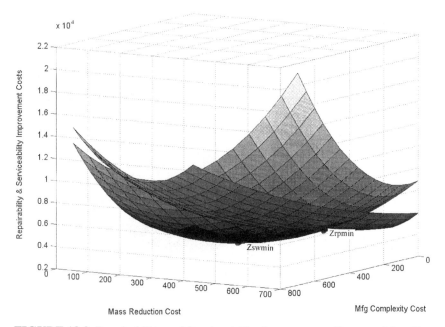

FIGURE 18.3 Repairability and Serviceability Improvement Costs vs. Mass Reduction and Manufacturing Complexity Costs

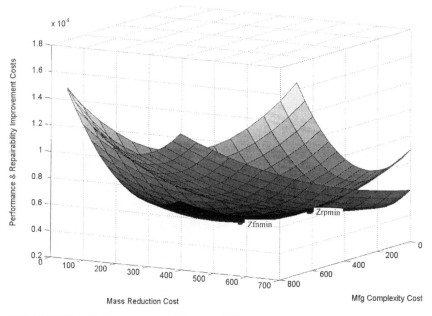

FIGURE 18.4 Performance and Repairability Improvement Costs vs. Mass Reduction and Manufacturing Complexity Costs

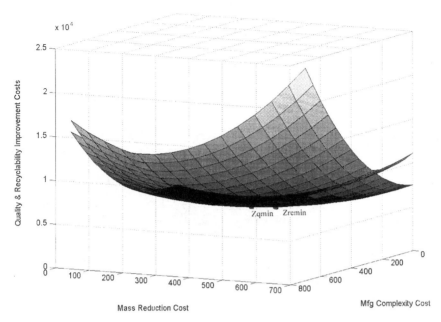

FIGURE 18.5 Quality and Recyclability Improvement Costs vs. Mass Reduction and Manufacturing Complexity Costs

White dots are needed to show minima when the shading contrast is too dark. Based on the curve shapes, it is not possible to show all eight minima on one plot. Combinations of four parameters are plotted together, in two separate graphs, while calculated values are shown in Table 18.1, as trials 41–48 for the eight parameters and 49 for the combined function of the 8 parameters, all weighted equally. (*Note*: trials 41-49 are the same as in Chapter 17.) Table 18.2 provides the key between the parameters and their MATLAB program designations.

Figure 18.1 shows safety and performance improvement cost trade-offs versus various combinations of mass reduction and manufacturing complexity costs. The trends shown indicate that any improvement in either parametric value will have to be "purchased" at great cost in mass reduction and complexity. Figures 18.2–18.5, for other pair of parameters, show the same trend, but not as dramatically as in Figure 18.1.

Figure 18.6 shows the first four of the eight design parameters plotted together. All four minima can be seen, with Zstmin and Zqmin close together. Figure 18.7 shows the other four parameters plotted together also

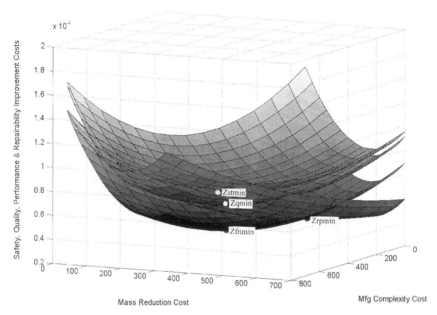

FIGURE 18.6 Safety, Quality, Performance, and Repairability Improvement Costs vs. Mass Reduction and Manufacturing Complexity Costs

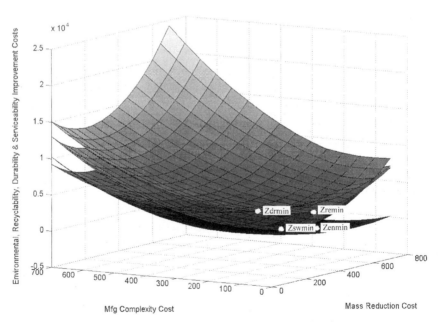

FIGURE 18.7 Environmental, Recyclability, Durability, and Serviceability Improvement Costs vs. Mass Reduction and Manufacturing Complexity Costs

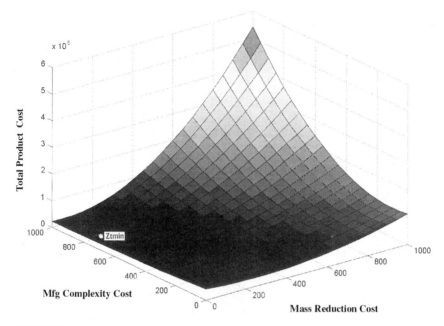

FIGURE 18.8 Total Product Cost vs. Mass Reduction and Manufacturing Complexity Costs

close together. Both plots confirm graphically the proximity of the minima produced analytically. These last two sets of curves also show the close proximity of the x and y minima to each other. Unequal weighting can be easily accomplished by first developing a multiplier based on the relative importance of the particular parameter in question, then applying it to the function, and finally applying it to the MATLAB *fmins* and *meshgrid* programs. Figure 18.8 shows the combined effects of all eight parameters minimized together.

TABLE 18.1. MATLAB Meshgrid Sample Command Sequence (for Four Surfaces: Heavy Duty Vehicles)

```
>> x_range=0:50:700;
>> y_range=0:50:700;
>> [X Y]=meshgrid(x_range,y_range);
>> Z1=(0.1.*(Y./2).^2-15.1.*(Y./2)+7900.9).*(X./600).^2-
   (0.1.*(Y./2).^2-7.5.*(Y./2)+5406.2).*(X./600)+
   0.1.*(Y./2).^2-7.6.*(Y./2)+7129.6;
>> Z2=(0.1.*(Y./2).^2-17.4.*(Y./2)+8373.3).*(X./600).^2-
   (0.1.*(Y./2).^2-6.2.*(Y./2)+4997.5).*(X./600)+
   0.1.*(Y./2).^2-4.2.*(Y./2)+5560.2;
>> Z3=(0.1.*(Y./2).^2-16.9.*(Y./2)+8297.9).*(X./600).^2-
   (0.1.*(Y./2).^2-5.5.*(Y./2)+4843.5).*(X./600)+
   0.1.*(Y./2).^2-3.6.*(Y./2)+3305.6;
>> Z4=(0.1.*(Y./2).^2-3.9.*(Y./2)+3946.2).*(X./600).^2-
   (0.1.*(Y./2).^2-3.7.*(Y./2)+6724.4).*(X./600)+
   0.1.*(Y./2).^2-17.8.*(Y./2)+8392.4;
>> surf(X,Y,Z1);
>> hold on
>> surf(x,Y,Z2);
>> surf(X,Y,Z3);
>> surf(X,Y,Z4);
>> Xlabel('Mass Reduction Cost');
>> Ylabel('Mfg Complexity Cost');
>> title('Safety, Quality, Performance, & Repairability Improvement
   Costs vs Mass Reduction & Mfg Complexity Costs');
>> Min_Z1=min(Z1(:));
>> Min_Z2=min(Z2(:));
>> Min_Z3=min(Z3(:));
>> Min_Z4=min(Z4(:));
>> [z1_row, z1_col]=find(Z1==Min_Z1);
>> [z2_row, z2_col]=find(Z2==Min_Z2);
>> [z3_row, z3_col]=find(Z3==Min_Z3);
>> [z4_row, z4_col]=find(Z4==Min_Z4);
>> min_x1=x_range(z1_col);min_y1=y_range(z1_row);
>> min_x2=x_range(z2_col);min_y2=y_range(z2_row);
>> min_x3=x_range(z3_col);min_y3=y_range(z3_row);
>> min_x4=x_range(z4_col);min_y4=y_range(z4_row);
>> plot3(min_x1,min_y1,Min_Z1,'go','MarkerFaceColor','g')
>> plot3(min_x2,min_y2,Min_Z2,'ro','MarkerFaceColor','r')
>> plot3(min_x3,min_y3,Min_Z3,'bo','MarkerFaceColor','b')
>> plot3(min_x4,min_y4,Min_Z4,'yo','MarkerFaceColor','y')
>> colormap gray
>> plotedit
>> ROTATE3d
```

TABLE 18.2. Index to MATLAB Functions Meshgrid & Fmins and Plot References

Dependent Parameter	Parameter Description	Function	Case	Trial No.	Figure No.	*fmins* Table No.
Zst	Safety	meshgrid	Heavy vehicle	41	18.1, 18.2, 18.6	18.1
		fmins	,,	,,		
Zq	Quality	meshgrid	,,	42	18.5, 18.6,	18.1
		fmins	,,	,,		
Zfn	Performance	meshgrid	,,	43	18.1, 18.4, 18.6	18.1
		fmins	,,	,,		
Zrp	Repairability	meshgrid	,,	44	18.2, 18.3, 18.4, 18.6	18.1
		fmins	,,	,,		
Zen	Environmental	meshgrid	,,	45	18.7	18.1
		fmins	,,	,,		
Zre	Recyclability	meshgrid	,,	46	18.5, 18.7	18.1
		fmins	,,	,,		
Zdr	Durability	meshgrid	,,	47	18.7	18.1
		fmins	,,	,,		
Zsw	Serviceability	meshgrid	,,	48	18.3, 18.7	18.1
		fmins	,,	,,		
Zt	Total product	meshgrid	,,	49	18.8	18.1
		fmins	,,	,,		

TABLE 18.3. MATLAB Fmins Program Example

When solving the fmins function, the first term after a = { } is Xabmin, the second is Yabmin. To find the equivalent Zabmin, locate Xmin and Ymin on the appropriate curve and read Zabmin on the vertical axis.

```
>> v=[0 0];
>> a=fmins('trial41_var',v)

a=

  Xmin=212.8569   Ymin=88.7013

>> v=[0 0];
>> a=fmins('trial42_var',v)

a =

  Xmin=183.9149   Ymin=49.9650

>> v=[0 0];
>> a=fmins('trial43_var',v)

a =

  Xmin=179.4646   Ymin=43.8646

>> v=[0 0];
>> a=fmins('trial44_var',v)

a =

  Xmin=482.3417   Ymin=205.9219

>> v=[0 0];
>> a=fmins('trial45_var',v)

a =

  Xmin=603.2212   Ymin=209.0138

>> v=[0 0];
>> a=fmins('trial46_var',v)

a =

  Xmin=260.4240   Ymin=4.4845

>> v=[0 0];
>> a=fmins('trial47_var',v)

a =

  Xmin=235.1697   Ymin=282.3069

>> v=[0 0];
>> a=fmins('trial48_var',v)

a =

  Xmin=168.6389   Ymin=55.2845

>> v=[0 0];
>> a=fmins('trial49_var',v)

a =

  Xmin=60.3835   Ymin=810.6265
```

19

ELECTRONIC PRODUCT MODEL DEVELOPMENT

19.1 INTRODUCTION: HOW PARAMETRIC ELECTRONIC SYSTEM DESIGN FUNCTIONS CAN BE OPTIMIZED

Chapter 14 illustrated the types of design problems and discussions of potential solutions. The fact that most of the issues were being discussed as early in the design phase as possible should not be interpreted to mean that the solutions were automatic. In most cases, a potential solution to one problem more than likely brought about several ancillary problems that also needed to be solved at the same time as the initial problem. Nonetheless, the design parameters were structured to address as many problems as possible for the first iteration, and then restructured to handle the others in later iterations.

19.2 PARAMETRIC DEFINITIONS AND TARGETS

The parameter target tasks are listed below:

Safety (Driver, Occupants, and Vehicle Integrity): Circuit breakers and fusable links to limit damage from electrical surges, short circuits, failed components, and damaging vehicle systems; glare shielding for heads-up display; etc.

Environmental Compliance: Shielding from electromagnetic interference; amortization of new circuit board production and electronic assembly equipment that meets anticipated new EPA and OSHA regulations.

Functional Performance: Expandable equipment bays, ports, and open connections for future expansion or additional memory chips, hard drive capacity, etc.

Durability/Reliability Improvement: Shock, temperature extreme, and "road" simulation test sampling for production units, prior to shipping.

Quality Improvement: Development of port replicators/docking stations to meet same requirements as the portable computer; simplified keyboard and voice actuation built in; improved battery life.

Repairability: Open circuit board modules for ease of removal and replacement, when damaged; improved diagnostics for internal self-checking for viruses and for program or hardware defects.

Recyclability: All plastic identified and recyclable; connectors and wiring field-strippable or demountable.

Serviceability: Modular, free-standing architecture allows for better isolation of problems for trouble shooting; improved written and electronic service documentation.

The parametric limits for the electronic products model were established similar to those for the automotive model with the smaller limits for mass reduction and complexity costs. The graphical solutions below provide for an optimized set of parameters, based on a generalized database of nominal costs.

19.3   GRAPHICAL SOLUTIONS

The plots show potential optimized solutions for several product design parametric costs for EMC's new portable PC. Figure 19.1 shows curves in descending order for safety, quality, durability, and serviceability improvement, with black or white dots as the minima. Table 19.1 shows the relationship between the MATLAB variables and the design parameters listed above. Table 19.2 lists all of the MATLAB *fmins* function outputs lists, following the key codes (i.e., Trial 111_var.m, and so on).

Similarly, other parametric combinations are plotted in Figures 19.3–19.5 so as to show minima not easily seen on previous plots or rotated for ease of analysis. The curves for each are generally in descending order (i.e., in the same order top to bottom as given in the title).

Since graphing all eight functions on the same plot would be too complex for analysis, the combined function is plotted in Figure 19.6, with

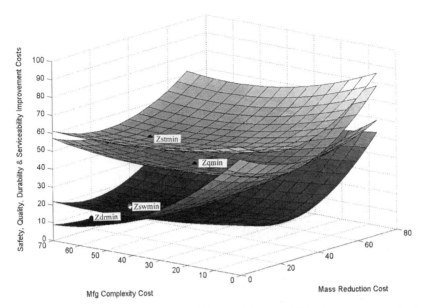

FIGURE 19.1 Safety, Quality, Durabilty and Serviceability Improvement Costs vs. Mass Reduction and Manufacturing Complexity Costs

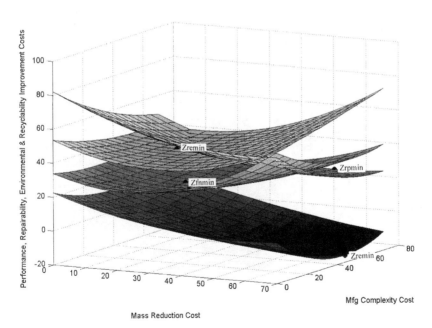

FIGURE 19.2 Performance, Repairability, Environmental, and Recyclability Improvement Costs vs. Mass Reduction and Manufacturing Complexity Costs

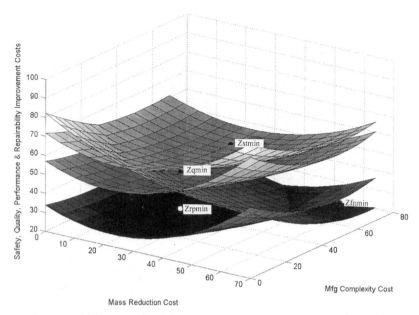

FIGURE 19.3 Safety, Quality, Performance, and Repairability Improvement Costs vs. Mass Reduction and Manufacturing Complexity Costs

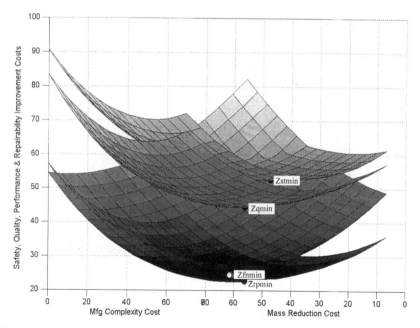

FIGURE 19.4 Safety, Quality, Performance, and Repairability Improvement Costs vs. Mass Reduction and Manufacturing Complexity Costs

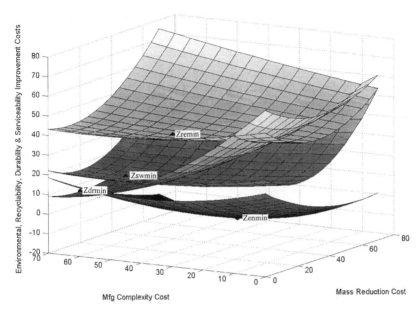

FIGURE 19.5 Environmental, Recyclability, Durability, and Serviceability Improvement Costs vs. Mass Reduction and Manufacturing Complexity Costs

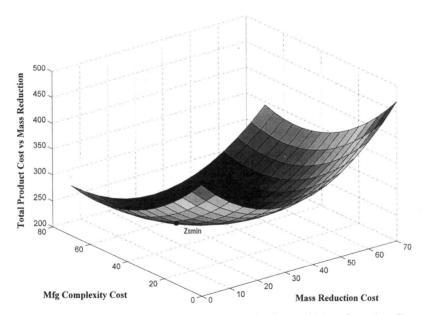

FIGURE 19.6 Total Product Cost vs. Mass Reduction and Manufacturing Complexity Costs

minimum values found from the *fmins* file. While this solution is not really optimized in the usual sense (i.e., it does not satisfy all conditions), at least it represents a compromise the meets all eight parameters as closely as possible. It was also assumed that the parameters had equal weight, an assumption which could be changed during the "what if" discussions that typically occur prior to new product design meetings. Given the normal tendency by some organizations to override technical issues with nontechnical ones, due to the complexity of the analysis, this format offers an opportunity to provide a more simplified approach to making intelligent new product decisions.

TABLE 19.1. MATLAB Meshgrid Sample Command Sequence (for Four Surfaces: Electronic Products)

```
>> x_range=0:50:700;
>> y_range=0:50:700;
>> [X Y]=meshgrid(x_range,y_range);
>> Z1=(.0293.*(Y./2).^2-1.6011.*(Y./2)+79.6473).*(X./75).^2-
   (.0131.*(Y./2).^2-.7433.*(Y./2)+54.0319).*(X./75)+
   .0149.*(Y./2).^2-.8178.*(Y./2)+71.6923;
>> Z2=(.028.*(Y./2).^2-1.8209.*(Y./2)+84.2846).*(X./75).^2-
   (.0112.*(Y./2).^2-.6582.*(Y./2)+50.2088).*(X./75)+
   .0175.*(Y./2).^2-.6022.*(Y./2)+56.8352;
>> Z3=(.0191.*(Y./2).^2-1.1982.*(Y./2)+52.6703).*(X./75).^2-
   (.0336.*(Y./2).^2-2.0472.*(Y./2)+50.0132).*(X./75)+
   .0422.*(Y./2).^2-3.0796.*(Y./2)+64.967;
>> Z4=(.0103.*(Y./2).^2-0.7011.*(Y./2)+84.9462).*(X./75).^2-
   (.0173.*(Y./2).^2-0.891.*(Y./2)+49.1308).*(X./75)+
   .0126.*(Y./2).^2-.6499.*(Y./2)+29.3462;
>> surf(X,Y,Z1);
>> hold on
>> surf(x,Y,Z2);
>> surf(X,Y,Z3);
>> surf(X,Y,Z4);
>> Xlabel('Mass Reduction Cost');
>> Ylabel('Mfg Complexity Cost');
>> title('Safety, Quality, Durability & Serviceability
   Improvement Costs vs Mass Reduction & Mfg Complexity Costs');
>> Min_Z1=min(Z1(:));
>> Min_Z2=min(Z2(:));
>> Min_Z3=min(Z3(:));
>> Min_Z4=min(Z4(:));
>> [z1_row, z1_col]=find(Z1==Min_Z1);
>> [z2_row, z2_col]=find(Z2==Min_Z2);
>> [z3_row, z3_col]=find(Z3==Min_Z3);
>> [z4_row, z4_col]=find(Z4==Min_Z4);
>> min_x1=x_range(z1_col);min_y1=y_range(z1_row);
>> min_x2=x_range(z2_col);min_y2=y_range(z2_row);
>> min_x3=x_range(z3_col);min_y3=y_range(z3_row);
>> min_x4=x_range(z4_col);min_y4=y_range(z4_row);
>> plot3(min_x1,min_y1,Min_Z1,'go','MarkerFaceColor','g')
>> plot3(min_x2,min_y2,Min_Z2,'ro','MarkerFaceColor','r')
>> plot3(min_x3,min_y3,Min_Z3,'bo','MarkerFaceColor','b')
>> plot3(min_x4,min_y4,Min_Z4,'yo','MarkerFaceColor','y')
>> colormap gray
>> plotedit
>> ROTATE3d
```

TABLE 19.2. Index to MATLAB Functions and Plot References

Dependent Parameter	Parameter Description	Function	Case	Trial No.	Figure No.	*fmins* Table No.
Zst	Safety	meshgrid	Electro	111	19.1, 19.3, 19.4	19.2
		fmins	"	111		
Zq	Quality	meshgrid	"	112	19.1, 19.3, 19.4	19.2
		fmins	"	112		
Zfn	Performance	meshgrid	"	113	19.2, 19.3, 19.4	19.2
		fmins	"	113		
Zrp	Repairability	meshgrid	"	114	19.2, 19.3, 19.4	19.2
		fmins	"	114		
Zen	Environmental	meshgrid	"	115	19.2, 19.5	19.2
		fmins	"	115		
Zre	Recyclability	meshgrid	"	116	19.2, 19.5	19.2
		fmins	"	116		
Zdr	Durability	meshgrid	"	117	19.1, 19.5	19.2
		fmins	"	117		
Zsw	Serviceability	meshgrid	"	118	19.1, 19.5	19.2
		fmins	"	118		
Zt	Total product	meshgrid	"	119	19.6	19.2
		fmins	"	119		

TABLE 19.3. MATLAB Fmins Program Example

When solving the fmins function, the first term after a = { } is Xabmin, the second is Yabmin. To find the equivalent Zabmin, locate Xmin and Ymin on the appropriate curve and read Zabmin on the vertical axis.

```
>> v=[0 0];
>> a=fmins('trial111_var',v)

a=

  Xmin=28.2404  Ymin=54.1673

>> v=[0 0];
>> a=fmins('trial112_var',v)

a =

  Xmin=25.8848  Ymin=34.8815

>> v=[0 0];
>> a=fmins('trial113_var',v)

a =

  Xmin=27.2016  Ymin=34.0204

>> v=[0 0];
>> a=fmins('trial114_var',v)

a =

  Xmin=53.4292  Ymin=77.9590

>> v=[0 0];
>> a=fmins('trial115_var',v)

a =

  Xmin=130.7010  Ymin=47.9611

>> v=[0 0];
>> a=fmins('trial116_var',v)

a =

  Xmin=18.0642  Ymin=41.9023

>> v=[0 0];
>> a=fmins('trial117_var',v)

a =

  Xmin=22.4051  Ymin=76.0315

>> v=[0 0];
>> a=fmins('trial118_var',v)

a =

  Xmin=19.1890  Ymin=52.8741
```

(continued)

TABLE 19.3. *(Continued)*

```
>> v=[0 0];
>> a=fmins('trial119_var',v)

a =

  Xmin=30.5018  Ymin=55.5225
```

Trial 111_var

```
function b=two_var(v);
X=v(1);
Y=v(2);
b=(.0293.*(Y./2).^2-1.6011.*(Y./2)+79.6473).*(X./75).^2-
  (.0131.*(Y./2).^2-.7433.*(Y./2)+54.0319).*(X./75)+
  .0149.*(Y./2).^2-.8178.*(Y./2)+71.6923;
%%%%%%%%%%%%%%%%%%%%%%%%%%%%%%end of function
```

Trial 112_var

```
function b=two_var(v);
X=v(1);
Y=v(2);
b=(.028.*(Y./2).^2-1.8209.*(Y./2)+84.2846).*(X./75).^2-
  (.0112.*(Y./2).^2-.6582.*(Y./2)+50.2088).*(X./75)+
  .0175.*(Y./2).^2-.6022.*(Y./2)+56.8352;
%%%%%%%%%%%%%%%%%%%%%%%%%%%%%%end of function
```

Trial 113_var

```
function b=two_var(v);
X=v(1);
Y=v(2);
b=(.0296.*(Y./2).^2-1.7144.*(Y./2)+79.2495).*(X./75).^2-
  (.0144.*(Y./2).^2-.6196.*(Y./2)+48.9176).*(X./75)+
  .0152.*(Y./2).^2-.4711.*(Y./2)+33.8396;
%%%%%%%%%%%%%%%%%%%%%%%%%%%%%%end of function
```

Trial 114_var

```
function b=two_var(v);
X=v(1);
Y=v(2);
b=(.0159.*(Y./2).^2-0.5153.*(Y./2)+40.3451).*(X./75).^2-
  (.0071.*(Y./2).^2-0.3799.*(Y./2)+67.3055).*(X./75)+
  .0165.*(Y./2).^2-1.5302.*(Y./2)+82.1802;
%%%%%%%%%%%%%%%%%%%%%%%%%%%%%%end of function
```

TABLE 19.3. (*Continued*)

Trial 115_var

```
function b=two_var(v);
X=v(1);
Y=v(2);
b=(.0289.*(Y./2).^2-1.4823.*(Y./2)+31.2055).*(X./75).^2-
    (.008.*(Y./2).^2-.4411.*(Y./2)+48.7725).*(X./75)+
    .0237.*(Y./2).^2-.9445.*(Y./2)+22.4714;
%%%%%%%%%%%%%%%%%%%%%%%%%%%%%%end of function
```

Trial 116_var

```
function b=two_var(v);
X=v(1);
Y=v(2);
b=(.0175.*(Y./2).^2-0.7952.*(Y./2)+65.6209).*(X./75).^2-
    (.0273.*(Y./2).^2-1.9988.*(Y./2)+57.1791).*(X./75)+
    .0134.*(Y./2).^2-.7638.*(Y./2)+53.5;
%%%%%%%%%%%%%%%%%%%%%%%%%%%%%%end of function
```

Trial 117_var

```
function b=two_var(v);
X=v(1);
Y=v(2);
b=(.0191.*(Y./2).^2-1.1982.*(Y./2)+52.6703).*(X./75).^2-
    (.0336.*(Y./2).^2-2.0472.*(Y./2)+50.0132).*(X./75)+
    .0422.*(Y./2).^2-3.0796.*(Y./2)+64.967;
%%%%%%%%%%%%%%%%%%%%%%%%%%%%%%end of function
```

Trial 118_var

```
function b=two_var(v);
X=v(1);
Y=v(2);
b=(.0103.*(Y./2).^2-0.7011.*(Y./2)+84.9462).*(X./75).^2-
    (.0173.*(Y./2).^2-0.891.*(Y./2)+49.1308).*(X./75)+
    .0126.*(Y./2).^2-.6499.*(Y./2)+29.3462;
%%%%%%%%%%%%%%%%%%%%%%%%%%%%%%end of function
```

(*continued*)

TABLE 19.3. (*Continued*)

Trial 119_var

```
function b=two_var(v);
X=v(1);
Y=v(2);
b=(.1786.*(Y./2).^2-9.8285.*(Y./2)+517.9694).*(X./75).^2-
   (.132.*(Y./2).^2-7.7791.*(Y./2)+425.5595).*(X./75)+
   .156.*(Y./2).^2-8.8591.*(Y./2)+414.8319;
%%%%%%%%%%%%%%%%%%%%%%%%%%%%%%%%%end of function
```

20
FUTURE DEVELOPMENTS

The science of Simultaneous Engineering is changing very rapidly. There were once discussions on how Simultaneous Engineering could provide benefits in real product design situations. Then came component, sub-system, and system optimization exercises, along with rapid prototyping through the utilization of stereolithography to produce prototype parts. Now, virtual reality design work stations have become a normal expectation. It is the adoption of these techniques that have allowed engineers to get "inside" their product and perfect it faster and more easily.

This treatise has attempted to promote the idea that, despite all this speed to market and rapid advance toward better-than-real-time engineering, there still needs to be a conscious effort to account for as many of the "outside" design influences and other potential product limitations as possible. To provide an all-electronic means for this activity to happen is perhaps wishful thinking and somewhat naive. For as much as an acceptable acceleration in design decision making is achieved, the responsibility for the decision also needs the same attention as the physical product design and development detail.

Based on current and hopefully not overly pretentious future trends, product liability, safety, security, and other issues will only assume larger proportions than now. Electronic decisions should not have the ability to instantly turn over an entire company's or industry's future, at least, not without that tiny pause to double-check that all of the nontechnical implications have been well thought out and agreed to by all concerned parties or their representatives. The consequences of this suggestion are not considered lightly. Competitive pressures, confidentiality issues, lead time to disseminate information, and financial implications are all extremely critical to any major new product development program.

This last reminder is to encourage the highest ethical, moral, and compassionate considerations when making new product decisions. The five industries presented herein represent a significant portion of the world's gross international product. Engineers and designers have the opportunity to advance civilization's output and outlook and should create the best environment for future generations as possible.

APPENDIXES

STRATEGIC PLANNING CHECKLIST

What is the average product life cycle in the company's industry or market area?

What new product decisions need to made that will generate the best results for the company?

What is the optimum timing for the new product introduction?

Will the company have the proper resource allocation to bring the new product to market on time and under budget?

Will the proper resources allow for simultaneous consideration of all engineering parameters?

Will the new product's financial performance contribute to the company's growth strategy?

How will the company's brand(s) be protected and enhanced by the new product?

MARKET RESEARCH CHECKLIST

Have the ideal customers been identified for the survey?

Does there exist an assessment of competitive products in the market?

Have all the target markets been identified and queried for their product needs and wants?

Have all current and competitive products been benchmarked for best design practices?

Have the Simultaneous Engineering parameters been considered in the survey questions?

Have adequate sample sizes been provided?

IDEA GENERATION CHECKLIST

Have the requirements for focus group membership been developed?

Do they include representation from all major customer groups?

Is the site of the idea generation session independent of regular workplace pressures and problems?

Will all of the current buying habits of your customers be represented on the focus group panel (i.e., those of the company's as well as of the competition)?

Have the buying decision factors for each product been ranked?

Are there any differences in perceived value between the company's product and the competition?

Have business buyers' concerns been addressed (e.g., existing product attributes, problems, price, delivery, warranty, operations within their systems, etc.)?

Has a list of the idea generation techniques been compiled?

Has a script been prepared for the focus group sessions?

Has a matrix been developed that compares the function, power requirements, and other features of the company's products with those of its competitors?

IDEA SCREENING AND EVALUATION CHECKLIST

What is the most likely reaction of the competition to the company's new product market entry?

Can competitive product features, costs, and prices be identified and ranked?

What is the ranking of the customer value of the new product compared with that of the competition?

What are potential substitute products for competitive offerings and their relative cost positions?

What will be the new product's expected sales growth, retail and wholesale price, production cost, overhead, investments, and cash flow. (First develop pro forma balance sheets, income statements, and cash flow analysis.) Utilize ratio analyses to compare new product to existing lines and competitors, if available.

What will be the breakeven volume for the first 5 years?

When will positive cash flow begin?

What percent market share is required to meet sales forecast?

What percent market shares do the top three competitors have in the same field as the new product?

What is the relationship between market-share target and economic performance and investment size and timing?

Has a risk analysis been performed to rank the new product relative to other potential investments for future changes in the market, economy, investment cost of capital? Has a worst-case scenario been considered?

Is the database in place to permit analysis of product sales performance?[1]

Have "what if," worst-case scenarios been considered?

PRODUCT PLANNING CHECKLIST

How will the new product meet the needs and wants of targeted customers?

Describe the major common characteristics of the buyer(s): Is there a significant difference between the targeted buyer of the new product and the end user?

What specific features and functions does the targeted customer value?

Describe the competitive advantage of the new product over the existing competition.

Rank the features and functions according to the buyers' purchasing criteria.

[1]Jeanemarie Caris McManus, *The New Product Development Planner*, AMACOM, a division of American Management Association, 135 West 50th Street, New York, NY 10020, 1991, pp. 63–68.

Are the features and functions unique to the new product?

Could the unique features and functions be easily copied by a competitor?

What could be done to prevent a competitor from copying the special features?

What are the costs associated with the differentiation process?

Can the most cost-effective or valuable differentiation activities be measured?[2]

Is the differentiation process sustainable in the perception of the buyer?

Can the new product be imitated?

How much would it cost a competitor to imitate the new product?

Is the new product an imitation?

Does the new product have features and benefits not currently available on existing products?

Does the competitor's cost structure permit a lower price to counteract the new product entry?

Can the new product price be based on value instead of cost?[3]

Can the differentiated value of the new product justify a premium price?

Could the new product price be too low?

Are the new product's competitors in a narrow or wide band of scope?

Is the new product's profitability based on a particular segment, or all segments?

What would be required to extend the new product's features into a new segment?

Will differentiation between segments require a new advertising and marketing strategy for each segment?

Is the new product concept adaptable to new customers, competitive situations, or outside influences (e.g., regulations, export requirements, etc.)

What areas of the new product have yet to be developed?

---

[2]McManus, pp. 57–61.
[3]McManus, pp. 85–92.

DESIGN AND ENGINEERING CHECKLISTS

Product Design Criteria

Is there a list of descriptions and specifications, of the product's functional (what it's supposed to do) and nonfunctional (appearance, styling, color) characteristics?

Is there a list of desirable outcomes, which might include safety, cost, packaging, power requirements, system compatibility, operational parameters, recyclability, etc.?

Automotive

Product Engineering Criteria

Compliance with safety standards.

Compliance with environmental regulations.

Manufacturability.

Warranty/serviceability.

Quality/customer satisfaction index.

Durability/reliability.

Performance/functionality.

Damageability/repairability/insurability.

Recyclability.

Future/unanticipated requirements.

Cost category analyses, within each of the above parameters, including development, opportunity, component, labor, and overhead.

Prioritization of the relationships between the above parameters and their singular and combined effects on the overall product.

Within the assembly process, include accessibility and disassembly for repair and material separation for recyclability, along with consideration for future legally binding regulations or future market opportunities.

Identification of all potential loss functions or hidden losses, quantified for elimination.

Design of experiments utilization to develop the interactions between potential design interactions.

Estimation of all performance parameters, including adequate power for passing acceleration and towing capacity, braking, lateral stability for maneuverability, power-train sizing, option content, and vehicle mass.

Regulatory parameter values (emissions and fuel economy standards), utilizing varying engine parameters such as displacement, valve timing, or fuel delivery to overcome a lack of performance.

Cost-of-ownership parameters such as depreciation, insurance, repair, operating, and maintenance: Damageability and repairability are the primary factors affecting insurance costs. Operating costs are related to fuel costs and availability, and maintainability factors are related to warranty experience and quality issues. Power train reliability is a direct contributor to maintenance costs.

Aerospace

Performance Characteristics to Be Considered for New Designs

Aerodynamics (lift, drag, moments, etc.)

Altitude (maximum ceiling and economic cruising range)

Speed (mach #, maximum, economic cruising, stall, sink, takeoff, and landing)

Range (miles or nautical miles, takeoff and landing distance)

Specific fuel consumption and capacity

Number of passengers and crew

Cargo capacity

Total gross weight (mass of aircraft, fuel, cargo and passengers)

Aircraft empty weight

Airfoil geometry and sizing (aspect ratio, wing sweep, taper ratio, twist, incidence, dihedral, vertical location, wingtips)

Airfoil lift and drag characteristics (design lift coefficient, stall, thickness ratio)

Reynolds number (air pressure and related forces)

Control surface sizing and balance

Tail geometry (arrangement, volume coefficient, spin recovery)

Thrust/weight (power loading, horsepower/weight, thrust matching)

Wing loading (stall speed, takeoff distance, landing distance, cruise, loitering, instantaneous and sustained turns, climb and glide, maximum ceiling)

Configuration layout and loft (wing/tail layout, wing location, wetted area determination)

Structural performance
Crashworthiness and survivability
Durability and reliability
Repairability and maintainability
Serviceability (accessibility)

Civilian Commercial Passenger Aircraft Profitability

Cost per seat-mile

Direct Costs:

Fuel
Crew Salaries
Maintenance
Depreciation (aircraft)
Insurance

Indirect or overhead costs:

Depreciation of ground facilities and equipment
Marketing expenses
Administration and general office overhead
Sales: Estimated on the type or class of seats and the number sold.
Load factor: The number of seats sold divided by the total available.
Revenue: per seat-mile is the average (i.e., coach) fare times the load factor.

Heavy/Military Vehicles

Functional Characteristics

Power plant and conversion
Track/wheel drive system
Payload type and capacity
Operational speed and range (with/without payload/equipment)
Operator safety and ergonomics
Instrumentation (control and display)

Stability requirements (center of gravity, with/without payload/equipment, outriggers, etc.)

Maneuverability (turning radius, approach and retreat angles)

Maintainability (serviceability, repairability)

Markets Segmentation/Applications

Soil, vegetation, and rock extraction and transport

Liquid cargo transport

Cranes and other high-lift operations

Firefighting and rescue

Medical transportation and injury evacuation

Refuse and recycling material removal and recovery

Passenger motor coach and recreational vehicle

Articulated tractor and trailer(s)

Electronic Products

In addition to the general product parameters listed above, the following are specifically related to electronic design:

Size

Geometry

Functional density

Event execution speed

Power dissipation

Parasitic effects

PROTOTYPE DEVELOPMENT CHECKLIST

The following is a Simultaneous Engineering checklist for prototype development:

Does the new product require a prototype or a limited production run to test modifications?

Are there technical complexities necessitating a prototype activity?

Is compliance testing sufficient for final design approval?

Is there functional equivalence to provide effective manufacturing feedback?

Will the prototype be sufficiently complete to provide the following?

  Marketing information
  Advertising and promotion
  Final cost structure and pricing policies
  Sales strategies
  Segmentation targeting
  Distribution channels
  Packaging

Specific Ground Vehicle Tests

Joint stiffness of both unitized and frame vehicle bodies
Front-end structure
Roof
Floor
Pillars
Underbody anchorages
Internal and external reinforcements

Analysis of Closures (Sealing Performance)

Doors
Hoods
Decklids

Other Performance Parameters

Vibration
Torsional rigidity
Crash energy absorption
Analysis of complete body and chassis bending and torsional rigidity
Durability/fatigue
Noise, vibration, and harshness (NVH)

Full Body/Chassis Fatigue

External sources

Road surface contacts through the tires

Wind noise

Aerodynamic loading of exterior panels and glass

Internal sources

Powerplant combustion reactions

Powerplant/power-train unbalance

Fuel supply slosh

Fan and compressor noise

Tire/wheel unbalance

Brake torque variation

Specific Aerospace Tests

Static and dynamic wing and control surface loading and flutter

Center of gravity and aerodynamic center measurement confirmation

Stall, spin, and recovery procedures

Maximum rate of climb, range, and altitude

Takeoff and landing distance

Heavy-Duty Vehicle Confirmation Tests

Electronic Product Tests

Drop shock

Vibration

Thermal shock

Temperature cycling

Salt fog

Humidity

Electrical shock

EMI

MANUFACTURING CHECKLIST

The following questions need to be asked before or during the manufacturing phase review unless already addressed in previous phases:

Resolution of all design-related issues

Resolution of all engineering questions

Resolution of all prototype feedback issues

Timing of tooling and facilities

Gauges and fixturing plan complete

Pilot or preproduction product plan

Part supplier concurrence

Inventory plan

Distribution plan

PROMOTION AND DISTRIBUTION CHECKLIST

Promotional Issues

New product analysis includes brand identification and valuation.

The promotion plan is considered and developed as part of the preliminary design proposal.

All potential media opportunities, including print, radio, television, billboard, and Internet, are considered in promotion.

The following need to be included in the approval of a new product's promotional plan:

Advertising agency selection

Cost (creative copy generation, media transformation)

Timing (frequency and market-segment exposure)

Media channel selection (print, radio, television, Internet, billboard, etc.)

Content approval (by corporate legal and brand management staff, language appropriateness, and clarity)

Trademark registration and copyright verification

Distribution Issues

Distribution plan considerations and development for the new product as part of the initial proposal

Packaging and warehousing considerations as part of the distribution plan

Special Shipping Requirements

Truck

Rail

Air freight

Labeling and bar coding requirements

REPAIRABILITY AND SERVICEABILITY CHECKLIST

Ground Vehicles

Design recommendations on optimizing damageability and repairability should consider the following basic premises concerning low-speed collision and theft damage:

Collision damage should be contained to the outboard areas of primary structural components (i.e., rails, pillars, for unibody vehicles, frame ends for [illegible] vehicles, etc.).

Repairs should be facilitated through ease of access or for removal and reinstallation of components.

Partial replacement procedures should be developed and available at the time of market introduction.

Replacement parts configuration should facilitate ease of installation.

Theft deterrence should include both content and drive-away countermeasures.[4,5,6]

Aerospace Vehicles

Design guidelines should consider the following as minimum requirements:

[4]*Optimizing Damageability, Repairability, Serviceability and Theft Deterrence*, Recommended Practice J1555 (Society of Automotive Engineers, Warrendale, Penn., 1985, rev. 1993), p. 1.

[5]*Vehicle Design Features for Optimum Low Speed Impact Performance* (Research Council for Automotive Repairs, 1995), pp. 1–3.

[6]Dieter Anselm, *Die Pkw-Karosserie Konstruktion, Deformationsverhalten, Unfallinstandsetzung*, (Vogel Verlag and Druk GmbH S Co., Wurzburg, Germany. 1997).

Provide accessibility to all electromechanical and hydraulic components and systems, having to remove other components and systems.

Provide electronically and print accessible diagnostic procedures for all electromechanical and hydraulic components and systems.

Provide repair kits for all electromechanical and hydraulic components and systems that are repairable.

Regular maintenance items should be grouped together to minimize vehicle operational interruptions for service.

Heavy-Duty Vehicles

Design guidelines should consider the following as minimum requirements:

Provide accessibility to all electromechanical and hydraulic components and systems, having to remove other components and systems.

Provide electronically and print accessible diagnostic procedures for all electromechanical and hydraulic components and systems.

Provide repair kits for all electromechanical and hydraulic components and systems that are repairable.

Regular maintenance items should be grouped together to minimize vehicle operational interruptions for service.

Electronic Products

Design guidelines should consider the following as minimum requirements:

Provide accessibility to all electronic components and systems, having to remove other components and systems.

Provide electronically and print accessible diagnostic procedures for all electronic components and systems.

Provide repair kits for all electronic components and systems that are repairable.

Regular software and hardware upgrade items should be grouped together to minimize vehicle operational interruptions.

SURVEY QUESTIONNAIRE

This questionnaire is to be filled out immediately after viewing the studio properties presented. If you have any questions while viewing the properties or afterwards, please contact the research clinic facilitators on the floor.

The eight properties that are shown represent several current production models, as well as new utility vehicle concepts, that may or may not reach production in the next few years. Please do not discuss any of these properties with anyone outside this facility or National Motors, Inc.

The first group of exhibits represent production vehicles that have been modified to hide their true identity without detracting from their utility characteristics. The facilitators will be loading and unloading the cargo areas during your review. They will ask you to assist them in these tasks, so that you may experience, firsthand, how the cargo capacity of each property can be maximized. The "cargo" consists of very light weight, imitation loads, which can be moved without any difficulty.

In later exhibits, the facilitators will ask you to assist them in converting some properties into alternate forms with greater or lesser utility than before. Please continue to assist the facilitators with their conversions, and loading and unloading exercises.

Finally, join us in the virtual laboratory to experience a driving simulation of these new concepts.

We know that you are very familiar with some of our current and competitive products. We would like your opinion on how well these new concepts meet your needs and wants regarding a new type of utility vehicle, one that can be readily adapted to your lifestyle requirements easily and quickly.

Please rank your satisfaction with the following properties, in the categories listed below:

How easy, practical, and convenient are the seat folding, storage, removal, and reinstallation procedures?

Property #	Completely Satisfactory	Partially Satisfactory	Neither	Partially Unsatisfactory	Completely Unsatisfactory
1	☐	☐	☐	☐	☐
2	☐	☐	☐	☐	☐
3	☐	☐	☐	☐	☐
4	☐	☐	☐	☐	☐
5	☐	☐	☐	☐	☐
6	☐	☐	☐	☐	☐
7	☐	☐	☐	☐	☐
8	☐	☐	☐	☐	☐

How easy, practical, and convenient are the latches, levers, and tie-downs for the roof racks, pods, roll-out containers, and tool carrier/workbench attachments?

Property #	Completely Satisfactory	Partially Satisfactory	Neither	Partially Unsatisfactory	Completely Unsatisfactory
1	☐	☐	☐	☐	☐
2	☐	☐	☐	☐	☐
3	☐	☐	☐	☐	☐
4	☐	☐	☐	☐	☐
5	☐	☐	☐	☐	☐
6	☐	☐	☐	☐	☐
7	☐	☐	☐	☐	☐
8	☐	☐	☐	☐	☐

How easy, practical and convenient are the optional storage modules for the rear interior cargo areas?

Property #	Completely Satisfactory	Partially Satisfactory	Neither	Partially Unsatisfactory	Completely Unsatisfactory
1	☐	☐	☐	☐	☐
2	☐	☐	☐	☐	☐
3	☐	☐	☐	☐	☐
4	☐	☐	☐	☐	☐
5	☐	☐	☐	☐	☐
6	☐	☐	☐	☐	☐
7	☐	☐	☐	☐	☐
8	☐	☐	☐	☐	☐

How easy, practical and convenient are the controls, handles, and mechanisms for operating the combination rear tailgate/liftgate?

Property #	Completely Satisfactory	Partially Satisfactory	Neither	Partially Unsatisfactory	Completely Unsatisfactory
1	☐	☐	☐	☐	☐
2	☐	☐	☐	☐	☐
3	☐	☐	☐	☐	☐
4	☐	☐	☐	☐	☐
5	☐	☐	☐	☐	☐
6	☐	☐	☐	☐	☐
7	☐	☐	☐	☐	☐
8	☐	☐	☐	☐	☐

How easy, practical, and convenient is the out-of-vehicle storage container for the roof pods, racks, tool, and roll-out containers?

Property #	Completely Satisfactory	Partially Satisfactory	Neither	Partially Unsatisfactory	Completely Unsatisfactory
1	☐	☐	☐	☐	☐
2	☐	☐	☐	☐	☐
3	☐	☐	☐	☐	☐
4	☐	☐	☐	☐	☐
5	☐	☐	☐	☐	☐
6	☐	☐	☐	☐	☐
7	☐	☐	☐	☐	☐
8	☐	☐	☐	☐	☐

Please use the space below to note any specific likes, dislikes, issues, and concerns not already covered:

_____

_____

_____

_____

This next series of questions is directed at the ease of vehicle entry and exit, seating comfort, seat and pedal adjustability range, seat belt fit and comfort, steering column adjustability range, windshield visibility, rear view mirror visibility and adjustability range, rear and side window visibility, instrument visibility, control movement efforts, and accessibility and convenience aspects of sitting in and operating these new utility vehicles.

Ease of vehicle entry and exit?

Property #	Completely Satisfactory	Partially Satisfactory	Neither	Partially Unsatisfactory	Completely Unsatisfactory
1	☐	☐	☐	☐	☐
2	☐	☐	☐	☐	☐
3	☐	☐	☐	☐	☐
4	☐	☐	☐	☐	☐
5	☐	☐	☐	☐	☐
6	☐	☐	☐	☐	☐
7	☐	☐	☐	☐	☐
8	☐	☐	☐	☐	☐

Seating comfort

Property #	Completely Satisfactory	Partially Satisfactory	Neither	Partially Unsatisfactory	Completely Unsatisfactory
1	☐	☐	☐	☐	☐
2	☐	☐	☐	☐	☐
3	☐	☐	☐	☐	☐
4	☐	☐	☐	☐	☐
5	☐	☐	☐	☐	☐
6	☐	☐	☐	☐	☐
7	☐	☐	☐	☐	☐
8	☐	☐	☐	☐	☐

Seat and pedal adjustability range

Concept #	Much Too Large	Too Large	About Right	Too Small	Much Too Small
1	☐	☐	☐	☐	☐
2	☐	☐	☐	☐	☐
3	☐	☐	☐	☐	☐
4	☐	☐	☐	☐	☐
5	☐	☐	☐	☐	☐
6	☐	☐	☐	☐	☐
7	☐	☐	☐	☐	☐
8	☐	☐	☐	☐	☐

Seat belt fit and comfort

Property #	Completely Satisfactory	Partially Satisfactory	Neither	Partially Unsatisfactory	Completely Unsatisfactory
1	☐	☐	☐	☐	☐
2	☐	☐	☐	☐	☐
3	☐	☐	☐	☐	☐
4	☐	☐	☐	☐	☐
5	☐	☐	☐	☐	☐
6	☐	☐	☐	☐	☐
7	☐	☐	☐	☐	☐
8	☐	☐	☐	☐	☐

Steering column adjustability range

Concept #	Much Too Large	Too Large	About Right	Too Small	Much Too Small
1	☐	☐	☐	☐	☐
2	☐	☐	☐	☐	☐
3	☐	☐	☐	☐	☐
4	☐	☐	☐	☐	☐
5	☐	☐	☐	☐	☐
6	☐	☐	☐	☐	☐
7	☐	☐	☐	☐	☐
8	☐	☐	☐	☐	☐

Windshield visibility

Property #	Completely Satisfactory	Partially Satisfactory	Neither	Partially Unsatisfactory	Completely Unsatisfactory
1	☐	☐	☐	☐	☐
2	☐	☐	☐	☐	☐
3	☐	☐	☐	☐	☐
4	☐	☐	☐	☐	☐
5	☐	☐	☐	☐	☐
6	☐	☐	☐	☐	☐
7	☐	☐	☐	☐	☐
8	☐	☐	☐	☐	☐

Rear and side window visibility

Property #	Completely Satisfactory	Partially Satisfactory	Neither	Partially Unsatisfactory	Completely Unsatisfactory
1	☐	☐	☐	☐	☐
2	☐	☐	☐	☐	☐
3	☐	☐	☐	☐	☐
4	☐	☐	☐	☐	☐
5	☐	☐	☐	☐	☐
6	☐	☐	☐	☐	☐
7	☐	☐	☐	☐	☐
8	☐	☐	☐	☐	☐

Rear view mirror visibility

Property #	Completely Satisfactory	Partially Satisfactory	Neither	Partially Unsatisfactory	Completely Unsatisfactory
1	☐	☐	☐	☐	☐
2	☐	☐	☐	☐	☐
3	☐	☐	☐	☐	☐
4	☐	☐	☐	☐	☐
5	☐	☐	☐	☐	☐
6	☐	☐	☐	☐	☐
7	☐	☐	☐	☐	☐
8	☐	☐	☐	☐	☐

Rear view mirror adjustability range

Concept #	Much Too Large	Too Large	About Right	Too Small	Much Too Small
1	☐	☐	☐	☐	☐
2	☐	☐	☐	☐	☐
3	☐	☐	☐	☐	☐
4	☐	☐	☐	☐	☐
5	☐	☐	☐	☐	☐
6	☐	☐	☐	☐	☐
7	☐	☐	☐	☐	☐
8	☐	☐	☐	☐	☐

Instrument visibility

Property #	Completely Satisfactory	Partially Satisfactory	Neither	Partially Unsatisfactory	Completely Unsatisfactory
1	☐	☐	☐	☐	☐
2	☐	☐	☐	☐	☐
3	☐	☐	☐	☐	☐
4	☐	☐	☐	☐	☐
5	☐	☐	☐	☐	☐
6	☐	☐	☐	☐	☐
7	☐	☐	☐	☐	☐
8	☐	☐	☐	☐	☐

Control accessibility

Property #	Completely Satisfactory	Partially Satisfactory	Neither	Partially Unsatisfactory	Completely Unsatisfactory
1	☐	☐	☐	☐	☐
2	☐	☐	☐	☐	☐
3	☐	☐	☐	☐	☐
4	☐	☐	☐	☐	☐
5	☐	☐	☐	☐	☐
6	☐	☐	☐	☐	☐
7	☐	☐	☐	☐	☐
8	☐	☐	☐	☐	☐

Control movement effort

Concept #	Much Too Large	Too Large	About Right	Too Small	Much Too Small
1	☐	☐	☐	☐	☐
2	☐	☐	☐	☐	☐
3	☐	☐	☐	☐	☐
4	☐	☐	☐	☐	☐
5	☐	☐	☐	☐	☐
6	☐	☐	☐	☐	☐
7	☐	☐	☐	☐	☐
8	☐	☐	☐	☐	☐

Overall convenience, fit, and comfort level

Property #	Completely Satisfactory	Partially Satisfactory	Neither	Partially Unsatisfactory	Completely Unsatisfactory
1	☐	☐	☐	☐	☐
2	☐	☐	☐	☐	☐
3	☐	☐	☐	☐	☐
4	☐	☐	☐	☐	☐
5	☐	☐	☐	☐	☐
6	☐	☐	☐	☐	☐
7	☐	☐	☐	☐	☐
8	☐	☐	☐	☐	☐

The final phase of the survey concerns the virtual driving simulator, which will help you experience what actually driving these new vehicles will feel like. You will don the virtual display helmet, and the facilitators will assist you with everything else. When you are ready, turn on the ignition switch and start driving! Before you do, remember that you will be controlling the vehicle, so pay close attention to traffic signals, other vehicles, and pedestrians. The exercise will be fun, but you must stay focused on the task before you and concentrate on driving fundamentals. Also, you will be going off-road, so expect some bumps, dips, dives, and other gyrations. Keep your seat belt tightly fastened, and let's go!

Please give us your impressions to these concept vehicles by answering the following questions:

How was the responsiveness of the vehicle in terms of available power, acceleration capabilities, load carrying, and trailer towing?

Property #	Completely Satisfactory	Partially Satisfactory	Neither	Partially Unsatisfactory	Completely Unsatisfactory
1	☐	☐	☐	☐	☐
2	☐	☐	☐	☐	☐
3	☐	☐	☐	☐	☐
4	☐	☐	☐	☐	☐
5	☐	☐	☐	☐	☐
6	☐	☐	☐	☐	☐
7	☐	☐	☐	☐	☐
8	☐	☐	☐	☐	☐

What about maneuverability, parking, passing in traffic, turning, backing up?

Property #	Completely Satisfactory	Partially Satisfactory	Neither	Partially Unsatisfactory	Completely Unsatisfactory
1	☐	☐	☐	☐	☐
2	☐	☐	☐	☐	☐
3	☐	☐	☐	☐	☐
4	☐	☐	☐	☐	☐
5	☐	☐	☐	☐	☐
6	☐	☐	☐	☐	☐
7	☐	☐	☐	☐	☐
8	☐	☐	☐	☐	☐

How stable is the vehicle on tight curves, on steep up and down grades, if off-road swales and gullies, when fording small streams, and climbing rocks, and loose gravel, mud, and snow?

Property #	Completely Satisfactory	Partially Satisfactory	Neither	Partially Unsatisfactory	Completely Unsatisfactory
1	☐	☐	☐	☐	☐
2	☐	☐	☐	☐	☐
3	☐	☐	☐	☐	☐
4	☐	☐	☐	☐	☐
5	☐	☐	☐	☐	☐
6	☐	☐	☐	☐	☐
7	☐	☐	☐	☐	☐
8	☐	☐	☐	☐	☐

How would you rate the vehicle on evasive maneuvering during unintentional skidding, on dry pavement and wet pavement, on snow, ice, loose gravel, and for collision avoidance?

Property #	Completely Satisfactory	Partially Satisfactory	Neither	Partially Unsatisfactory	Completely Unsatisfactory
1	☐	☐	☐	☐	☐
2	☐	☐	☐	☐	☐
3	☐	☐	☐	☐	☐
4	☐	☐	☐	☐	☐
5	☐	☐	☐	☐	☐
6	☐	☐	☐	☐	☐
7	☐	☐	☐	☐	☐
8	☐	☐	☐	☐	☐

How effective is the antilock braking and traction control system compared with other similarly equipped cars that you have driven under similar conditions?

Concept #	Much Better Than Average	Better Than Average	Average	Worse Than Average	Much Worse Than Average
1	☐	☐	☐	☐	☐
2	☐	☐	☐	☐	☐
3	☐	☐	☐	☐	☐
4	☐	☐	☐	☐	☐
5	☐	☐	☐	☐	☐
6	☐	☐	☐	☐	☐
7	☐	☐	☐	☐	☐
8	☐	☐	☐	☐	☐

Concept #	Much Too Large	Too Large	About Right	Too Small	Much Too Small
1	☐	☐	☐	☐	☐
2	☐	☐	☐	☐	☐
3	☐	☐	☐	☐	☐
4	☐	☐	☐	☐	☐
5	☐	☐	☐	☐	☐
6	☐	☐	☐	☐	☐
7	☐	☐	☐	☐	☐
8	☐	☐	☐	☐	☐

How is the cargo convertibility compared with other similarly equipped vehicles?

Concept #	Much Better Than Average	Better Than Average	Average	Worse Than Average	Much Worse Than Average
1	☐	☐	☐	☐	☐
2	☐	☐	☐	☐	☐
3	☐	☐	☐	☐	☐
4	☐	☐	☐	☐	☐
5	☐	☐	☐	☐	☐
6	☐	☐	☐	☐	☐
7	☐	☐	☐	☐	☐
8	☐	☐	☐	☐	☐

Cargo capacity

Concept #	Much Too Large	Too Large	About Right	Too Small	Much Too Small
1	☐	☐	☐	☐	☐
2	☐	☐	☐	☐	☐
3	☐	☐	☐	☐	☐
4	☐	☐	☐	☐	☐
5	☐	☐	☐	☐	☐
6	☐	☐	☐	☐	☐
7	☐	☐	☐	☐	☐
8	☐	☐	☐	☐	☐

Seating convertibility

Concept #	Much Better Than Average	Better Than Average	Average	Worse Than Average	Much Worse Than Average
1	☐	☐	☐	☐	☐
2	☐	☐	☐	☐	☐
3	☐	☐	☐	☐	☐
4	☐	☐	☐	☐	☐
5	☐	☐	☐	☐	☐
6	☐	☐	☐	☐	☐
7	☐	☐	☐	☐	☐
8	☐	☐	☐	☐	☐

Ease of ingress/egress

Concept #	Much Better Than Average	Better Than Average	Average	Worse Than Average	Much Worse Than Average
1	☐	☐	☐	☐	☐
2	☐	☐	☐	☐	☐
3	☐	☐	☐	☐	☐
4	☐	☐	☐	☐	☐
5	☐	☐	☐	☐	☐
6	☐	☐	☐	☐	☐
7	☐	☐	☐	☐	☐
8	☐	☐	☐	☐	☐

Overall, how satisfied were you with these concepts?

Property #	Completely Satisfactory	Partially Satisfactory	Neither	Partially Unsatisfactory	Completely Unsatisfactory
1	☐	☐	☐	☐	☐
2	☐	☐	☐	☐	☐
3	☐	☐	☐	☐	☐
4	☐	☐	☐	☐	☐
5	☐	☐	☐	☐	☐
6	☐	☐	☐	☐	☐
7	☐	☐	☐	☐	☐
8	☐	☐	☐	☐	☐

PRODUCT PLANNING PROPOSAL FORM

NATIONAL MOTORS CORPORATION

PRODUCT PLANNING PROPOSAL FORM: NEW VEHICLE PROGRAM/PLATFORM
PROGRAM/PLATFORM DESCRIPTION:
  PROGRAM HIGHLIGHTS:
    MAKE/MODEL/YEAR & JOB #1 TIMING:
      MAKE: MODEL:
      MODEL YEAR: JOB #1 TIMING:
      PLATFORM DERIVATION: ALL NEW: MODIFIED C/O:
      RESKIN: FRESHEN:
    REF: STUDIO PROPERTY #/ CHASSIS MOCKUP #:
    DRIVELINE FEATURES:
      POWER UNIT:
          TRANSFER UNIT(S):
          FINAL DRIVE UNIT(S):
    CHASSIS FEATURES:
      SEPARATE FRAME:
      STEERING:
      SUSPENSION:
      BRAKES:
      TIRES & WHEELS:
    BODY & TRIM FEATURES:
      BODY PANELS:
      PRIMARY STRUCTURE:
      SECONDARY STRUCTURE
      MOVABLE SUBASSEMBLIES:
      BUMPER SYSTEM PERFORMANCE & COMPONENTS:
    PAINT, COLOR, TRIM LEVELS & CONTENT:
      BASE:
      SECOND LEVEL:
      THIRD LEVEL:
    MARKET PERFORMANCE EXPECTATIONS:
      MARKET SEGMENT:
      NEW ENTRY/REPLACEMENT OF:
      MARKET SHARE POTENTIAL:
      CANNIBALIZATION POTENTIAL:
      ANTICIPATED COMPETITIVE REACTION(S):
FINANCIAL PERFORMANCE EXPECTATIONS:
  VARIABLE PROFIT:
  BREAKEVEN VOLUME:
  INCREMENTAL FIXED COSTS:
    HUMAN RESOURCE/STAFFING:
    PRODUCT DEVELOPMENT:
    FACILITIES ALLOCATION:
  RETURN ON INCREMENTAL INVESTMENT:
DESIGN/ENGINEERING/MANUFACTURING/SERVICE EXPECTATIONS:
  DESIGN EXPECTATIONS:
  ENGINEERING EXPECTATIONS:
  PROCUREMENT EXPECTATIONS
  MANUFACTURING EXPECTATIONS:
  SERVICE EXPECTATIONS:

OTHER EXPECTATIONS: (DISTRIBUTION CHANNEL, ADVERTISING,
   INSURABILITY, WARRANTY, RESIDUAL VALUE AFTER LEASE, ETC.):

                APPROVAL/DISAPPROVAL   DATE   COMMENTS ATTACHED

FINANCE:
RESEARCH:
LEGAL:
PRODUCT PLANNING:
DESIGN:
ENGINEERING:
MANUFACTURING:
SERVICE:
DEALER COUNCIL:
ADVERTISING AGENCY:
SUPPLIER GROUP:
MARKETING:
CORPORATE:

GENERAL REFERENCES

Anselm, Dieter, *Die Pkw-Karosserie, Konstruktion, Deformationsverhalten, Unfallinstandsetzung*, Vogel Verlag und Druck GmbH & Co. KG Wurzburg, Germany, 1997. English translation by Ian Findlay, in process, 1999.

Handbook of Business Strategy, Faulkner & Gray, New York, 1996.

Jackson, Paul, et al., *Janes's All The World's Aircraft*, Jane's Information Group, Alexandria, Va., 1996.

Kelly, Orr, *King of the Killing Zone*, W. W. Norton, New York, 1989.

Mendenhall, William, and Richard L. Schaefer, *Mathematical Statistics with Applications*, Wadsworth Publishing Co., Belmont, Calif., 1973.

Miller, Landon C. G., *Concurrent Engineering Design*, Society of Manufacturing Engineers, Dearborn, Mich., 1993.

Ohtsuki, T., series ed., *Advances in CAD for VLSI*, Elsevier Science Publishers, Amsterdam, 1986.

Raymer, Daniel P., *Aircraft Design: A Conceptual Approach*, American Institute of Aeronautics and Astronautics, Washington, D.C., 1992.

RCAR, *Design Features for Optimizing Vehicle Damageability*.

RCAR, *Design Guideline for Vehicle Security*.

RCAR, *The Procedure for Conducting a Low Speed 15km/h Offset Insurance Crash Test to Determine the Damageability and Repairability Features of Motor Vehicles*.

Reinhard, K., *Introduction To Integrated Circuit Engineering*, Houghton Mifflin, Boston, Mass., 1987.

SAE Standard J326 Nomenclature—Hydraulic Backhoes.

SAE Standard J727 Nomenclature—Crawler Tractor.

Note: SAE documents are available from Society of Automotive Engineers International, 400 Commonwealth Drive, Warrendale, PA 15096-0001.

RCAR documents are available from RCAR (Research Council for Automobile Repairs), Stallarholmsvagen 31, Box 90, S-12421, Stockholm, Sweden.

317

SAE Standard J729 Nomenclature and Specification Dimensions—Dozers.

SAE Standard J1057 Identification Terminology of Earthmoving Machines.

SAE Standard J1193 Nomenclature and Dimensions for Hydraulic Excavators.

SAE Standard J1234 Specification Definitions—Off-Road Work Machines.

SAE Aerospace Recommended Practice ARP 4104—Design Objectives for Handling Qualities of Transport Aircraft.

SAE Recommended Practice J1555—Optimizing Automobile Damageability.

INDEX

319